BOTANIC
GARDENS
OF THE
WORLD

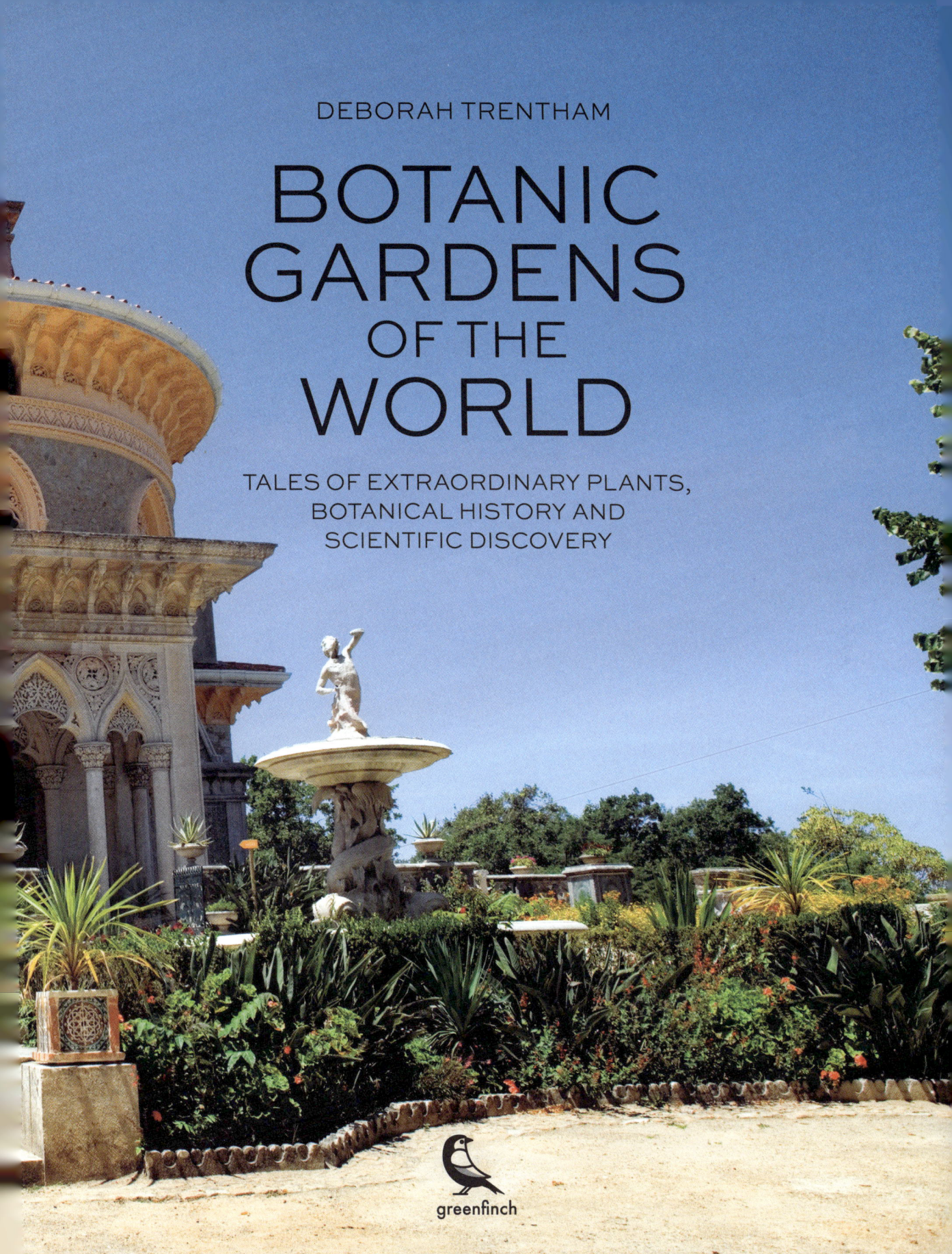

DEBORAH TRENTHAM

BOTANIC GARDENS OF THE WORLD

TALES OF EXTRAORDINARY PLANTS, BOTANICAL HISTORY AND SCIENTIFIC DISCOVERY

greenfinch

Contents

Introduction

Designed as spaces of study and research, botanic gardens are where the lavish beauty of the natural world collides with cutting-edge scientific developments. Throughout history the role of botanic gardens has shifted according to the context of the time, from primarily places of medical research to key theatres in the age of exploration, and more recently with a focus on conservation in the face of the escalating climate challenges.

With their earliest origins stretching back to the ancient Levant, it is said that there are currently over 2,500 botanic gardens in existence around the world. Today, they are forging the way in conservation efforts and scientific research, but the roots of these horticultural institutions are much older. From early blueprints such as the 4th century BC garden of Aristotle, or the oldest botanic gardens that have been in continual use – the Garden of the University of Padua, established in 1545. In this book we explore the history of botanic gardens through 40 examples of the most important and fascinating gardens of the world. Beginning with the earliest examples, exploring the early origins of botanic gardens in Renaissance Italy at Pisa and Padua, then moving through the centuries to the present day and discovering how the creation and purpose of botanic gardens has evolved throughout history.

LEFT: Mosaïcultures Internationales. In 2013 the competition took place in Montréal Botanic Gardens. Mosaïculture is the art of making three-dimensional sculptures from plants and flowers, here over three million flowers were used to produce sculptures which were then displayed in the garden.

OPPOSITE: The sub-tropical gardens at Tresco Abbey on the Isles of Scilly have their own microclimate.

The botanic gardens chosen are not always the biggest, most famous or most preserved, some are chosen for their historic values, some because they are pushing ahead with scientific research, others are repositories for rare plant species and those facing extinction. The Botanic Garden Conservation International stated in 2018 that the criteria to define a botanic garden were to 'have a greater emphasis on conserving rare and threatened plants, compliance with international policies, and sustainability and ethical initiatives.'

Medicinal Botanicals

Since the beginning of time people have strived to find healing properties in plants and to identify plants with medicinal qualities. In Ancient Egypt we know plants were used for medicine and collected by royalty. One of the oldest pieces of evidence is the Ebers Papyrus (now in the library of the University of Leipzig), a medical papyrus scroll written circa 1550 BC. It reveals how ancient Egyptians used and collected plants, containing 800 prescriptions referring to 700 plant species and drugs used for healing. Many of them are plants we recognize today, such as pomegranate, garlic, onion, fig, and coriander. We also know that pharaohs sent parties out searching for plants: one was Queen Hatshepsut, who around 1500 BC sent five ships to the Land of Punt to bring back living specimens of plants and desert trees for her.

Evidence in ancient texts tells us that plants were always considered for their utility, it is only much later that we find the study of pure botany and applied botany. The work of Pliny the Elder (born in 23 AD) is all lost apart from one survival, *Historia Naturalis*: an epic undertaking of 37 volumes, it is an encyclopaedia of everything to do with the world. It was copied many times but is considered the largest Roman work to come down to us. Amongst the numerous subjects he tackles are cosmology, astronomy and for us of

DIOSC ORIDE

Chi più intese di me quel che Natura
Celò nell' herbe, e le virtù secrete:
Arriuò il mio saper all' alte mete,
Où huom mortale in, uan gionger procura.

in Bassano per il Remondini

LIBER PRIMVS.

Ita. Angoloſo odorato gionco.

Gal.Souchet.

Cyperus, quem aliqui Eryſiſceptrum, ut aſpalathum appellant, folia fert porraceis proxima, ſed longiora & exiliora: caulem odorati iunci, cubitale aut maiorem, anguloſum, in cuius acumine minuta folia cū ſemine exoriuntur: radices, quarum in medicina eſt uſus, cohærentes inter ſe, oleis oblongis ſimiles, aut rotunda, nigro colore, ſuaui odore, amaro guſtu. In lacunoſis, paluſtribus & cultis enaſcitur. Optima eſt grauiſſima, denſa, fractu contumax, plena & aſpera, odoris cum quadam acrimonia iucundi, qualis eſt Cilicia, Syriaca, & quæ à Cycladibus inſulis petitur. Huic calida uis ineſt. Venarum ſpiracula laxat, urinam ciet, ad calculos, aquamq́ ſubtercutem utiliſſime bibitur, aduerſus ſcorpionum ictus remedio eſt: perfrictionibus uuluæ, & præcluſionibus fotu prodeſt: pellit menſes. Contra ulcera oris, etiam ſi depaſcant uorentq́, farina eius efficax eſt. Malagmatis calfacientibus, & unguentorum ſpiſſamentis bene adijci ſolet. Aliud Cyperi genus tradunt in India naſci, gingiberis effigie, quod manducatum croci uim reddit, guſtu amaro ſentitur, & illitum præſentem pſilothri uim.

Cyperus

particular interest, botany, agriculture, horticulture and pharmacology. In Book XX Pliny describes kitchen garden plants, giving the medicinal use for each of them: for example, he says if chicory is boiled it 'loosens the bowels and benefits the liver'. In Book XXII he tells us he does his own research into the medicinal use of herbs because he doesn't trust doctors, who he says don't properly investigate the herbs and medicines they prescribe.

One of the early Greek botanists was Theophrastus, whose seminal works include *Enquiry into Plants* and *On the Causes of Plants* which classified plants and explained the economics of growing them. Another Greek writer, Pedanius Dioscorides, wrote the classic work of ancient times *De Materia Medica* in circa 77 AD. This text is incredibly important as it references botanic tradition, it survived in various forms, it was translated into Latin from Greek in late antiquity more than once, and was one of the first scientific texts to have a significant effect on Arabic medicine as well as medical practices across Europe. It gives the information on medicinal plants that was to constitute the basic *materia medica* until the late Middle Ages and the Renaissance. The Vienna Dioscorides is the most beautiful and the oldest manuscript copy to survive. It was made 512 AD in Constantinople for a Byzantine princess, Anicia Juliana. It contains hundreds of exquisite images of plants and animals, some but not all depicted naturalistically. Pliny the Elder wouldn't have approved as he said it was inadvisable to illustrate

ABOVE: An illustrated page from an early printed version of Dioscorides' *De Materia Medica*, dated 1552.

OPPOSITE: A portrait of the Greek physician, pharmacologist and botanist, Pedanius Dioscorides, by Remondini.

RIGHT: A beautifully illustrated folio from an Arabic translation of Dioscorides' *De Materia Medica*, possibly from Baghdad.

herbal texts as inaccuracy could be dangerous. He wasn't wrong to be wary, Theophrastus was a philosopher and botanist, Dioscorides was a physician, pharmacologist and botanist but not all the early authors came with this background. Other writers were not botanists and were unfamiliar with many of the plants they were documenting.

The Age of Discovery

Important factors that impacted plant movement across the globe included, from the 14th century onwards, Portuguese sailors and their voyages of discovery, followed swiftly by the Spanish and other Europeans. Although it was the later 15th century Portuguese discoverers who filled the royal courts in Lisbon with gold, spices, illuminated manuscripts, talking birds, exotic animals and plants. The gardens and menageries were lost in the Lisbon earthquake and tsunami of 1755, as was the royal library with its manuscripts and its attached Cabinet of Natural History. In 1453, two major world events happened: one was the invention of the printing press, and Pliny's *Historia Naturalis* was published in Venice in 1469. The other event was when the Ottoman army captured Constantinople, one outcome of this was the circulation of Greek scientific treatises and texts into western Europe.

However the botanic gardens as we know them today began in the 16th century in Italy with the advent of iatrochemistry. The Italian Renaissance saw the medicinal simples gardens of the monasteries evolve into physic gardens, these were attached to centres of learning and were the forerunners of today's botanic gardens. The gardens of Pisa and Padua were the first to be founded for the academic study of medicinal plants, Universities across Europe were quick to follow and gardens such as Leiden and Oxford were founded. They were still using the ancient texts, including those of Theophrastus and Dioscorides, while at the same time new plants were being brought into Europe from other continents, Africa and the Americas, these were plants which people were not sure how to use, whether they were medicinal or culinary. The rich trading ports of Venice and the Amalfi coast brought not only spices and treasures into Italy but also these exotic and unknown plants. New classifications and investigations were needed to understand how the plants could be utilized.

Some of these new and exotic finds were going into the hands of private collectors, the Renaissance Humanist garden makers inspired by classical literature, such as Homer (who mentions hundreds of different plants) and the gardens described in the letters of Pliny

LEFT: The *Cantino Planisphere* of 1502 shows the world as known by the Portuguese after their great explorations at the end of the 15th century.

The Age of the Enlightenment saw the explorers looking for scientific discoveries around the world, using new technological innovations and taking with them naturalists and botanists. The Western royal courts were the great patrons, the Kings of England and France sending many on voyages of discovery. Later, as empire took hold and the western explorers travelled to more remote places, the plant hunters went too. For the colonizers wanted crops and slaves to grow them and whole agricultural histories would be wiped out for new crops to be grown that would be profitable for the colonizers. The plant hunters had their own lucrative business, bringing rare, curious and exotic plants to the excited (and rich) garden owner.

Kew Gardens and other gardens, such as *Real Jardín Botánico de Madrid* were fundamental to the plant hunting and discoveries that took place during the 18th and 19th centuries. They were focused on cultivating the new species that were being brought back from expeditions. Some botanic colonial gardens were formed with the directives from Kew Gardens, others independently but often with economic botany as well as the horticulture and science at their core.

The Future of Botanical Gardens

Today the role of the botanic garden has changed. Some act as repositories for the rare specimens, to protect the world's plant diversity threatened with extinction, others are more focused on *ex situ* conservation to save plants. The botanic gardens set up through colonies and

the Younger and in the 'Metamorphoses' of Ovid. The Italian courts were vying with each other to have the best in art, architecture and gardens. Families such as the Medici in Tuscany were instrumental in the forming of botanic gardens in Pisa and in Florence, these were status symbols but they also became repositories of scientific knowledge about nature and the world. In their private gardens the Renaissance nobility had the best and rarest plants they could get their hands on. The Gonzagas in their Ducal Palace in Mantua had a *Giardino dei Semplici* – a botanical garden and a hanging garden of orange trees. The d'Estes of Ferrara with their 'Delizie' with rare plants and exotic animals, all were plant collectors.

ABOVE: The Royal Botanic Gardens, Kew in London were central to the discovery of plant species in the 18th and 19th centuries.

empires are now making the break from centuries of oppression. They are now investigating and establishing the endemic plants that were erased from their natural habitat. Some were lost through the exploitation of the land, crops brought from other lands to be grown cheaply and at great cost to the inhabitants whose territories (and they themselves) were used for profit. Illegal plant trade is a huge problem, as is unsustainable plant collecting. The poaching of plants puts more pressure on ecosystems, removing plants that are contributing to the native habitat (i.e. plants that prevent erosion and are then removed) and endangering plants into extinction.

Botanic gardens are investigating and researching plant taxonomy and genetics, phytochemistry, hoping to find useful plant properties for global issues such as climate change. The threat of climate change must be one of the biggest challenges and plants are being studied to see which can adapt and withstand the environmental damage. Education remains one of the main aims of botanic gardens as it was in the beginning, they now need to teach us how to care and hopefully mend the world we humans have been damaging.

Orto Botanico di Pisa
Italy (1544)

Minutes away from the crowds of tourists who flock to the Piazza dei Miracoli, the oldest University Botanical Garden in the world is hidden behind high walls. Its calm serenity is in marked contrast to the piazza, where masses of people with phones and cameras congregate to photograph themselves with the Leaning Tower of Pisa.

History

The garden at Pisa was established in 1543–1544 by Cosimo I de' Medici, Duke of Florence, and later Grand Duke of Tuscany (in 1569). He created not one but two botanical gardens: a year after Pisa he founded a second in Florence. Cosimo was a keen plant collector: he already had gardens full of rare and expensive exotics, but it seems he was not the pioneer he appears to be. The garden at Pisa was developed primarily to encourage Luca Ghini to take up the post of Cosimo's private physician. Ghini was Professor of Medicine at Bologna University; his efforts to establish a teaching garden for his students there had come to nothing, and as a condition of accepting the position offered to him in Pisa, he insisted that Cosimo should pay for a private garden of medicinal plants. He took the title of 'Professor of simples' (*Professor Simplicium*), and a 'garden of simples' (physic garden) was built as a teaching aid for his students. It was the first of its kind, and was to be copied by universities around Europe. Ghini completed its construction, and assembled a collection of many rare and interesting plants. He is credited with

RIGHT: Pisa, the *Orto Botanico*. The building seen to the right is the *Palazzo delle Conchiglie* (originally the Botanic Institute), whose rocaille shellwork was added in 1752.

inventing the herbarium, a collection of dried plant specimens that could be studied at any season of the year, which he brought with him to Pisa. During his time at the garden he was responsible for teaching many influential botanists, including Andrea Cesalpino, who was to succeed him as superintendent of the garden (1554–1558).

Originally built on the right bank of the Arno River, it was moved closer to the University in 1563, and in 1591 was relocated to its present location near the Piazza dei Miracoli. Over time, more land was acquired for it, and it now occupies 3 hectares (7.4 acres).

The giraffes

Perhaps the most unusual addition to the garden over the centuries happened in the 1800s, when it became home to a family of giraffes. A drawing of a single animal in a circular enclosure was made (probably from a hot-air balloon) by French artist Alfred Guesdon in around 1849, and recent research by Gianni Bedini and Simone Farina at Pisa University has revealed correspondence from Paolo Savi, then the garden's director, which mentions that the giraffe, 'a male, now lively and cheerful, so that he seems very healthy...has a cow with calf in his company.'

The seven sectors of the garden

The Botanical Garden is organized into seven areas, each holds collections which are arranged scientifically, and there are over 6,000 plants from five continents, including a range of succulents from the desert regions of Africa, America and Mexico, Mediterranean flora, and native species from the nearby Tuscan marshes.

The *Orto del Cedro* (Cedar Garden) has seven monumental trees, including an Oriental Plane (*Platanus orientalis*) planted in 1808 that measures over 25m (80ft), while the *Orto del Gratta* has a pond of aquatic plants, some of which are endangered in the wild.

The *Orto del Mirto* (Myrtle Garden) takes its name from the myrtle tree (*Myrtus communis*) planted here in

1815. It features about 140 species of medicinal plants (reflecting the collection of the original Garden of Simples), and is regularly used by Pharmacy students from Pisa University.

The Botanical Museum (*Scuola Botanica*)

The Botanical Museum, founded in 1591 by Grand Duke Ferdinando I de' Medici, is now situated in the '*Palazzo delle Conchiglie*' – the 'Palace of Shells', named for its 18th-century façade decorated with seashells and coral. Inside is a gallery of portraits of famous botanists with links to the garden, a wonderful 16th-century Tuscan oak doorway that was once the entrance to the gardens, and the recreated Cabinet of Curiosities or *Wunderkammer*. Early visitors recorded they saw whale bones, a stuffed crocodile and all kinds of strange fossils and animals.

The objects in the museum are mainly related to the teaching of botany. One room is full of very rare, early botanical models of citrus and fungi, created in beeswax and plaster, made by Luigi Calamai and his pupils in Florence between 1830–1840. There are over one hundred accurate models: the wax citrus were records of the fruits grown, but the fungi models were used to identify and teach edible and poisonous varieties. There are also reproductions of various botanical plant sections in wax, enlarged for teaching purposes. The *Herbarium Horti Botanici Pisani* is held here and can be viewed by appointment: it contains over 350,000 specimens that have been gradually amassed here since the end of the 18th century.

Glasshouses (*Serre*)

One of the directors of the garden, Michelangelo Tilli (1655–1740) introduced heated rooms for growing exotic plants at the garden; he was the one of the first botanists to attempt this. The Banana Greenhouse was completed in the 19th century: it is the oldest greenhouse in the garden. There are three others, including a small one, the Victoria Greenhouse, containing a pool of water lilies (*Victoria cruziana*) from the Amazon river.

Piazzale Arcangeli

In 1890, the director, Giovanni Arcangeli, planted two Chilean wine palms in a newly landscaped area of the garden. In 2018, one of the palms was attacked by the invasive red palm weevil *Rhynchophorus ferrugineus*: native to southern Asia, it came into Europe by accident in the 1990s. This 'concealed tissue borer' inflicted so much damage on the palm that it had to be cut down in 2020; it has been replaced by a young seedling, grown from seeds of the original twin palms.

Teaching remains one of the fundamental aspects here and the Garden opens the world of plants to all, including tours designed for the non-hearing, people with Alzheimer's and children affected by Spectral Autism.

And one last thing in the arboretum – there are trees dating from hundreds of years ago, including a rare Chinese ginkgo planted at the end of the 18th century...and a view of the Leaning Tower!

Orto Botanico di Padova
Padua, Italy (1545)

The *Orto Botanico* of Padua is the oldest University Botanical Garden in the world that has been in continual use. Unlike the other gardens, this one has never moved, and is listed on the UNESCO World Heritage site.

Padua University, known as one of the greatest seats of learning for medicine and anatomy, is the second-oldest university in Italy, founded in 1222. (The University of Bologna, dating from 1088, is believed to be the world's oldest.) Astronomer Nicolaus Copernicus (1473–1543) studied at Padua, and Galileo Galilei (1564–1642) was its professor of mathematics from 1592 to 1610.

At the beginning of the Renaissance, plants were not an independent subject of enquiry, but were only seen in terms of their possible practical value, especially for pharmacological purposes.

By the end of the 16th century, *giardini dei semplici* (simples gardens) would be replaced with gardens where plants could be studied: the *horti botanici*. A new way of seeing and interpreting the natural world was now emerging, no longer through folklore and superstition, but through modern science.

Padua, as a University city in the trading centre of the rich Veneto region, was ideally placed for a key role in this. Spices and medicinal plants from the first voyages of discovery were arriving in Venice, and needed to be dealt with properly. The necessity for medicines had been the reason for the formation of the earliest botanic gardens in Europe, which had been located in monasteries and religious communities.

Plants as medicine

As early as 1533 the *Universitas Artistarum* of Padua founded the first chair of *materia medica*, the branch of medical science concerned with the study of drugs used in the treatment of disease. As time went on, the identification of the flora described by ancient authors became a critical issue. The early manuscript books containing their details had often been recopied inaccurately: they sometimes lacked illustrations, or even contained images that no longer resembled the original plants, causing mistakes and confusions that led to the deaths of patients.

In 1545, the Senate of the Venetian Republic recognized that an increased knowledge of herbal remedies would reduce errors in pharmacy; and, following the advice of botanist Francesco Bonafede (1474–1558), it was decided to found a garden specifically for cultivation and research of medicinal plants. This was to be the Botanic Garden (*Orto Botanico*) of Padua.

The science of Pharmacognosy

Bonafede was the first to teach Pharmacognosy as a natural science instead of only using classical texts: the definition of Pharmacognosy is knowledge of medicinal drugs obtained from plants or other natural sources. Bonafede's students had the plants whose uses he was describing physically in front of them as he taught them, thereby avoiding previous confusions. His approach attracted wide attention: academics and students came to Padua

A World Heritage Listed Garden

'The Botanical Garden of Padua is the original of all botanical gardens throughout the world, and represents the birth of science, of scientific exchanges, and understanding of the relationship between nature and culture. It has made a profound contribution to the development of many modern scientific disciplines, notably botany, medicine, chemistry, ecology, and pharmacy.' –UNESCO

ABOVE: The glasshouse on the right was built for the oldest plant in the *Orto Botanico,* dating from 1585, known as Goethe's Palm (*Chamaerops humilis*).

from across Europe, and its Garden became the point of reference for the founding of similar *horti botanici*.

The scientific approach in ordering collections of plants – in all their forms, living, dried and illustrated – was highly influential, and led to a change of interest in plants, which were now to be studied not just for their healing properties, but as an independent subject matter: pure botany.

Designing the Botanical Garden of Padua

A responsibility of every botanical garden is to collect a great variety of plants, and organize them into categories and species, with the aim of allowing their observation and study. The Botanical Garden of Padua holds 3,500 different species, representing (though on a reduced scale) a significant part of the entire Plant Kingdom.

The first keeper of the Garden in 1546 was Luigi Squalermo (1512–1570). It is still in its initial location and has maintained its original layout more or less unchanged. The land on which it stands once belonged to a Benedictine monastery of Santa Giustina (where there would have been an area cultivating plants for medicines). The identity of the designer of the Garden is still debated but may well have been Daniele Barbaro (1514–1570), a Venetian nobleman, learned humanist, and a patron of architect Andrea Palladio.

Margaret Muther D'Evelyn says he had 'overseen the design of the artistically arranged Botanic Garden in Padua', and the architect Andrea Moroni

ABOVE: An early 16th-century plan of the circular garden layout of Padua by Girolamo Porro, 1591. The garden keeps this original design, although a later addition was the circular enclosing wall, built to protect the plants from the frequent night thefts.

from Bergamo was certainly involved in its construction.

The Garden follows the rules of harmony and proportion as used by Palladio, consisting of a circle (84 metres/275 feet in diameter) enclosing a square, which is then divided into four by two crossing pathways.

The four squares, called 'quarters', 'tiers' or 'terraces' because they were

ABOVE: Originally the Garden at Padua contained medicinal plants with the emphasis on teaching students from the University, but soon the collection swelled with the arrival of exotics and rare plants. Coming from all over the world, some were brought in by traders of the Venetian Empire.

night – the plants were rare, and the medicines that could be made from them were very valuable. A few years later the stone balustrade and tall gates were also added for security.

The plants

At the beginning the Garden held only medicinal plants, but with the arrival of flora from all over the world, brought in by traders of the Venetian Empire, it was soon boasting a wide range of exotics and rarities.

originally raised about 70 centimetres (27.5 inches) higher than the paths, were divided into beds which were arranged to create geometrical patterns. The shapes of the beds were used as a kind of *aide memoire* for the students: intricate and complex, but designs that could be recalled.

In 1552 a circular enclosing wall was built, as there were constant thefts at

They came from various places – through plant-finding expeditions or via the trade routes. Bulbs arrived from Turkey, other specimens from the New

OPPOSITE AND ABOVE: In 2014 a new wing of the Botanical Garden at Padua was unveiled, called the 'Garden of Biodiversity', designed by Giorgio Strappazzon and VG Associati. One of the most advanced greenhouses in the world, it contains more than 1,300 species of plants collected from every climatic zone of the Earth.

World via Portugal. Some were received as gifts from aristocratic families of the region – or were sent to Italy as a result of the connections made with naturalists in other gardens.

The Garden's development was supported by the gradual establishment of a herbarium, a library, and a number of laboratories which continue to be added to and developed. In 2014 a new glass extension was opened, designed by Giorgio Strappazzon: this Biodiversity Garden consists of five greenhouses that each hold plants from a different continent.

Hortus Botanicus Leiden
The Netherlands (1590)

In the southwest of Leiden's ancient city centre is the *Hortus botanicus*, the first botanic garden in the Netherlands. Founded in 1590, three years after the city's University requested permission from its mayor to establish a physic garden for the benefit of medical students, it drew inspiration from the botanic gardens in Pisa and Padua. Its creation was entrusted to the celebrated French botanist, Carolus Clusius. Now containing more than 10,000 species, the garden has justifiably been described as a 'living museum.' Committed to teaching, research and the conservation of at-risk species, the *Hortus* works closely with, among others, Leiden's acclaimed Naturalis Biodiversity Center.

RIGHT: A view of the *Hortus Botanicus Leiden* showing the modern reconstruction of the early Clusius garden of 1594. Originally the garden was intended for educational purposes and contained mainly medicinal plants.

BELOW: A colour engraving of the *Hortus Botanicus* of the University of Leiden; the bird's-eye view reveals the planting beds and layout of the garden. The Image dates from 1610.

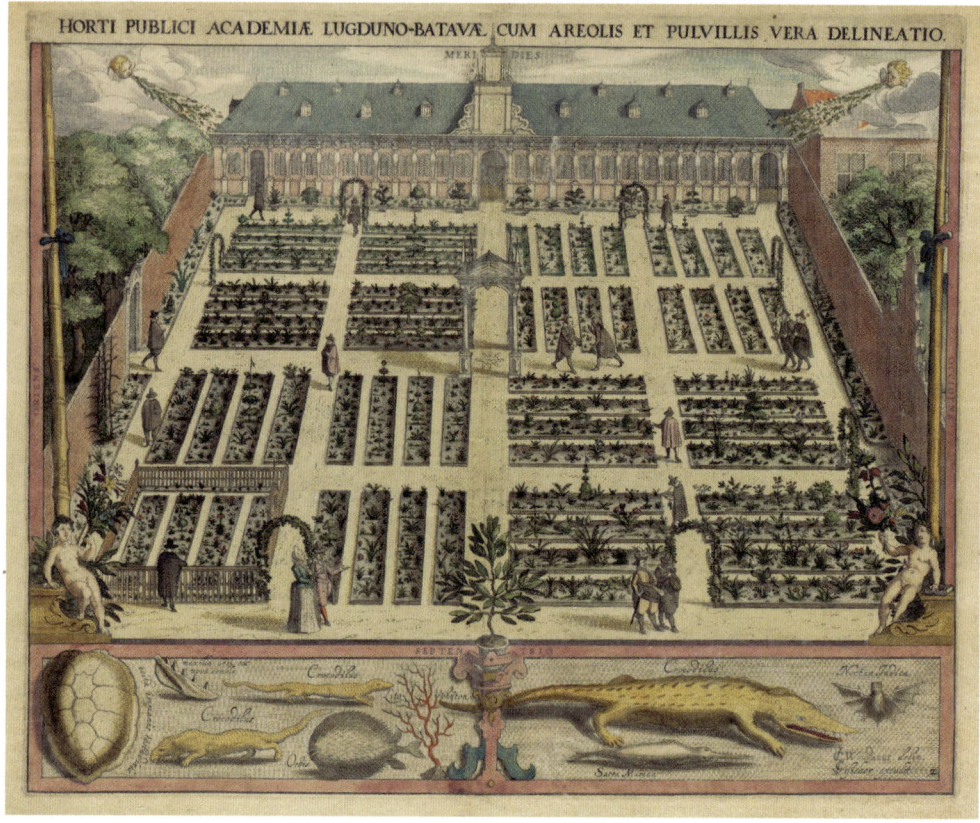

Carolus Clusius

Clusius – also known as Charles de l'Écluse – was born in Arras, France, in 1526, and was well into his sixties before he set to work in Leiden. His appointment came after he had spent sixteen years establishing a medical garden in Vienna for the Holy Roman Emperor, Maximilian II. Leiden benefitted from Clusius' years of extensive travel, which included a detailed exploration of the Iberian Peninsula resulting in the introduction of some 200 new plants to Northern Europe.

Moreover, he could call upon an unrivalled network of botanic experts and contacts, numbering at least 300. The expertise of this web of correspondents took in not only Europe but Asia and the New World, opening up new realms of botanic specimens. It was, for instance, at a meeting with Sir Francis Drake that Clusius was introduced to both cocoa and the sweet potato. Among other exotic species he cultivated were the chestnut, narcissus, hyacinth and, perhaps most famously, the tulip.

Leiden's proximity to ports used by the Dutch East India Company (*Vereenigde Oostindische Compagnie*) was also advantageous. Clusius persuaded its directors to authorize the collection of plants for him, so providing a plethora of species from Dutch trading colonies in India and across South Asia.

Although the *Hortus* at Leiden initially had a relatively small footprint, Clusius crammed in over a thousand species. Moreover, while Leiden was envisaged as a physic garden, he put it at the forefront of the development of botany not merely as a sub-strand of medicine but as an independent scientific discipline in its own right.

Tulips

Leiden's collection of medicinal plant species was quickly augmented by a tropical collection that included some selected simply for their exotic beauty. Over the years, the *Hortus* became synonymous with the tulip, which had recently been introduced into Europe from Turkey. Clusius' observations – not least his recognition that certain colour variations were the result of infection by the 'tulip breaking virus' – deepened understanding of the species, and were instrumental in the founding of the Dutch bulb industry. However, he kept some of the most impressive specimens in his own personal garden, fearing that in the *Hortus* they would become prey to the epidemic of exotic species thefts then plaguing gardens across the continent.

The Glasshouses

Glasshouses were essential for the survival of exotic species in Northern Europe, and the first were built in the late 17th century. The Leiden *Hortus*'s Orangery went up in the 1740s. At the heart of the garden's present-day glasshouse complex is the Victoria Glasshouse, dating from the early 1800s.

RIGHT: The Victoria glasshouse in the *Hortus Botanicus*, built for the giant waterlily, *Victoria amazonica*. It first bloomed here in 1872, and continues to do so.

Elegantly constructed from cast iron and with a tented roof, it famously houses the giant *Victoria amazonica* waterlilies, discovered in the then British colony of British Guiana (now Guyana) in 1801 and later named for Queen Victoria. Botanic gardens competed to see whose specimens would flower earliest – and the first time the ones in Leiden bloomed in 1872, more than 30,000 visitors came to gaze upon them. Even today, parents can arrange to pose for photos with their babies on the plant's giant leaves.

Adjoining the Victoria Glasshouse is the Orchid House, whose core collection derives from South-East Asia. The complex also boasts two glasshouses with treetop canopy walks, as well as a distinctive elevated classroom balcony. Until 1930, the houses relied on coal for heating, with members of staff stoking the fires 24 hours a day. Later, oil proved ferociously expensive during the energy crisis of the 1970s, and since 1983, gas has provided the heating.

The *Hortus* Today

The *Hortus* has continuously expanded and evolved over the centuries, and the modern Front Garden – divided into four quadrants – sits on the site of Clusius' original plot, hosting a historical reconstruction using plants from his own lists.

Alongside a Fern Garden and a Systematic Garden (which aims to showcase the major groups of the plant kingdom over 32 beds), there is a Japanese Garden commemorating the botanic and wider cultural exchanges between Japan and the Netherlands since

a Dutch ship initiated first contact in 1600. It also houses the Von Siebold Garden, in honour of Philipp Franz von Siebold (1796–1866) the German botanist who sent a great many plant species to Leiden from his travels in Japan. Its plants and tea room are protected from the elements by a bright red wall inspired by the tea houses of Nagasaki, where Dutch traders once found their rest and relaxation.

ABOVE: Historic greenhouse in the *Hortus Botanicus*. Glasshouses were important from the early days of the garden to care for plants that began to arrive from hotter climes.

The Chinese Herb Garden

Among the more recent additions to the *Hortus* is a Chinese herb garden, constructed in 2015 and a vital resource for scholars at Leiden University investigating the history of Chinese medicinal plants, their modern impact and potential future uses. In co-operation with the China Academy of Chinese Medical Sciences in Beijing, academics are engaged in ongoing, groundbreaking chemical analysis and DNA barcoding of the garden's plants.

Jardin des Plantes de Montpellier
France (1593)

Established in 1593 by Henri IV, the Garden was originally intended for the pharmacological education of student doctors and apothecaries at the city's celebrated university. Lost and remade more than once over the centuries, that does not diminish the importance of either the site or the vision of those who have moulded it. Still faithful to its early purpose as a teaching garden, it contains some 2,680 species, including almost 500 from the Mediterranean, along with collections of medicinal and subtropical specimens. With its fresh air and sea breezes, the garden's 4.6 hectares (11.3 acres) are today a Historical Monument and Protected Site.

History

Montpellier's university, founded in 1220, is one of the oldest in the world. In an age when most medicines were based on herbal preparations, its status as a busy port city with a constant inflow of medicines and spices from around the world attracted scholars from across Europe.

King Henri IV brought in the French doctor and botanist Pierre Richer de Belleval (1564–1632) as the Garden's designer and director. It was based on the models of Pisa/Padua, laid out with beds organized into plant families. The original layout consisted of the King's Garden (containing medicinal plants), the Queen's Garden (with alpine plants, notably from Languedoc) and the King's Square (for plants of particular botanical, as opposed to medicinal interest). It soon built impressive collections, not only

RIGHT: A view of a walkway in the *Jardin des Plantes de Montpellier*.

of indigenous species but from around the Mediterranean, from more distant tropical climes and from the Baltic as well.

Like Pisa and Padua, it had an academic role, with lectures on botany and *hortus* plant demonstrations where students would learn to identify the plants, and became the model for the French botanical garden (including the *Jardin des Plantes* established over forty years later in Paris). Montpellier's reputation soon spread, with the 17th-century Danish travel writer,

Peter Eisenberg, writing in 1614 in his *Itinerarium Galliæ et Angliæ* (*Travels in France and England*): '...from the outside [it] does not seem lavish, but it is rich and excellent, with lots of plants, some of them bizarre.'

The Destruction of the Garden

Montpellier was a Protestant Huguenot stronghold but in 1622, Louis XIII's chief adviser, the Catholic Cardinal Richelieu, sent troops in a bid to impose control. To fend off impending attack, the citizens of Montpellier built up the city's fortifications, with the Garden chosen as the site of one of the first bastions. As Charles Frédéric Martins (the Garden's director from 1851–1880 – see below) reported in 1852:

> When the siege of Montpellier, in 1622, came to destroy the fruits of so many pains and labours. Like a tender father who rushes into danger to save his children, Richer de Belleval removed the most precious plants from the Garden and transported them to that of the School of Medicine, belonging currently at the School of Pharmacy.

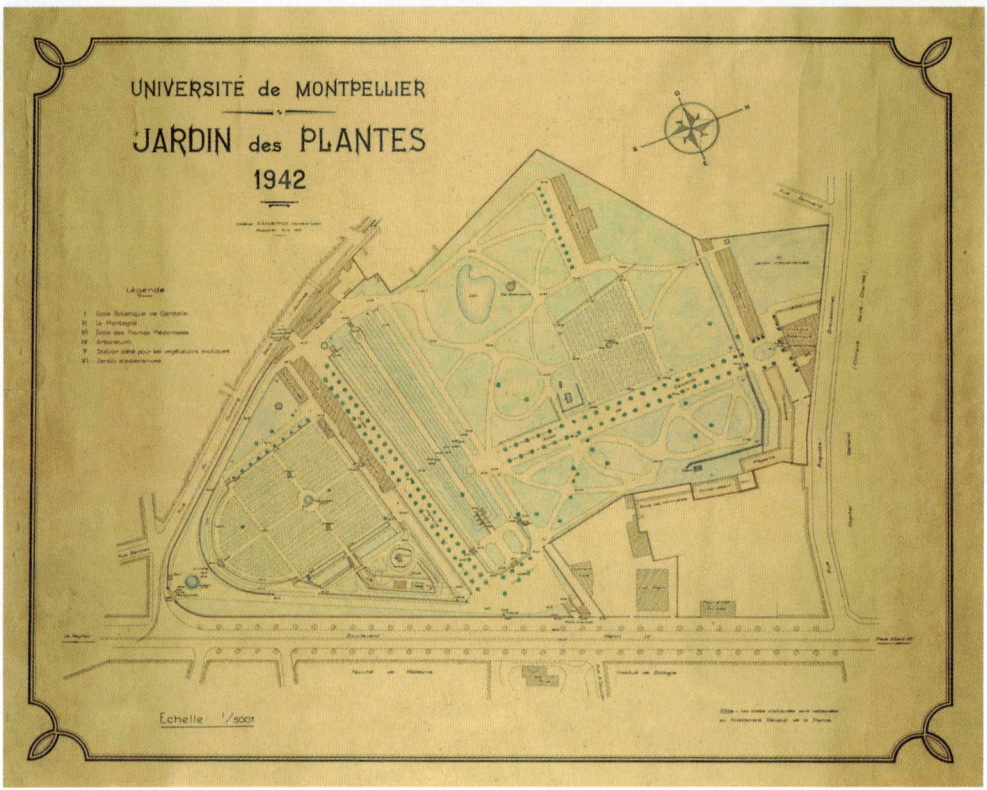

Despite Richer de Belleval's efforts, it seems that most of the Garden's treasures were lost during this episode, and he spent the rest of his life attempting to rebuild it from scratch, a job that was carried on after his death by his nephew. The Garden must have been rebuilt to some degree by 1629 as Cardinal Richelieu visited the Garden in June taking with him his whole court.

Belleval's Mountain

Among Richer de Belleval's innovations was what he called the *Monticule*, which remains today as a series of beds radiating down a slope that contain plants from all types of terrain from alpine to coastal, and from shade-seeking to sun-loving. The amphitheatre-style terracing ensures there is a consistently wonderful display of plants.

Another famous 17th-century director was Pierre Magnol, who developed his Systematic School and published the first classification of plants by family. The father of modern taxonomy, Carl Linnaeus, later named a genus of Magnolia in honour of him.

The Orangery and Glasshouse

François Boissier de Sauvages de Lacroix, a former student at the university, went on to hold the chair of botany and constructed the Garden's first greenhouse in the 1740s. Montpellier's orangery, designed by Claude-Mathieu Delagardette, opened in 1804. It was the pet project of Pierre Marie Auguste Broussonet, a native of the city who was made director of the garden in 1803, having previously fled France during the Revolution. Broussonet's tenure saw a

OPPOSITE: This plan of the Garden dated 1942 by Alfred Ansermoz was commissioned by the University of Montpellier. It shows the Institute of Botany on its original site to the north before it was replaced in the 1950s.

ABOVE: The restored Orangerie at the *Jardin des Plantes* originally dates from 1806.

Make a Wish

Among the host of Montpellier's remarkable trees is a *Ginkgo biloba* (Chinese maidenhair tree) planted in 1788 by the French naturalist and pioneer of Linnaean taxonomy Antoine Gouan. Said to be the first of the species grown in France, it was presented to Gouan by Broussonet, who in turn had been given it by the celebrated English naturalist, Sir Joseph Banks. However, the oldest extant tree at Montpellier is an incredible 400-year-old *Phillyrea latifolia* (Oleaceae). Known as the 'wishing tree', it's customary for people to write their secret heart's desires on small notes which they hide within the tree.

LEFT: Dome of the 1879 Observatory within the garden, originally fitted with the highest quality Foucault telescope.

RIGHT: The Martins greenhouse of 1860 created by Charles Frédéric Martins. Radically altered in the 1950s, it has now been restored to its original design and houses cacti and succulents.

major expansion of the collections and a sensitive development of the site.

In 1860, the garden's then director, botanist and zoologist Charles Frédéric Martins, created the large greenhouse that now bears his name and which has been restored in recent years. Today, it houses over 420 species over three rooms (a tropical room with a large central basin for aquatic plants, and two temperate rooms). Among its many points of interest are a world-leading collection of specimens from French Guiana.

International Influences

The English Garden, with its pond and glasshouse, was originally landscaped in the English style back in 1859. The arboretum is older still, dating to 1810, and today contains 570 tree species. Other highlights include a lotus pool (the *lac aux Nélombos*) and a tropical greenhouse with more than 420 species, including orchids, *Bromeliaceae* and a wonderful collection of palms, including *Brahea armata* and *Butia capitata*.

Botanisk Have
Copenhagen, Denmark (1600)

The University of Copenhagen Botanical Garden is in the city centre, part of the *Parkmuseerne* museum district. Established in 1600 by Royal Charter as a *Hortus Medicus* (medical garden), the Garden that we see today, covering some 10 hectares (24.7 acres), dates from 1870 and is the fourth iteration of the botanical garden. It comprises Denmark's largest collection of living plants, with over 13,000 species, including 600 native ones, 1,100 perennials and 1,100 annuals. Highlights include a Conifer Hill and an important collection of carnivorous plants, as well as rock gardens providing habitat for plants indigenous to the more mountainous regions of Central and Southern Europe. Famed for its Victorian-era glasshouses, the Garden is administratively part of the Natural History Museum of Denmark under the aegis of Copenhagen University's Faculty of Science.

History

The original *Hortus Medicus* stood on land donated by King Christian IV, probably to house plants from the gardens of religious houses abandoned during the Reformation. However, with no financial support the Garden suffered and was not maintained. In 1620 it fell under the control of Ole Worm (Olaus Wormius), a polymath who was professor of medicine at the University of Copenhagen. He promoted its use in teaching medical students, and expanded the collection both with native plants and rare species sent from his contacts abroad.

OPPOSITE: An aerial view of Copenhagen Botanical Garden, which is located in the very centre of the city.

BELOW: Plan of the Copenhagen Botanical Garden (*Botanisk Have*), dated 1907. The Palm House can be seen to the northwest.

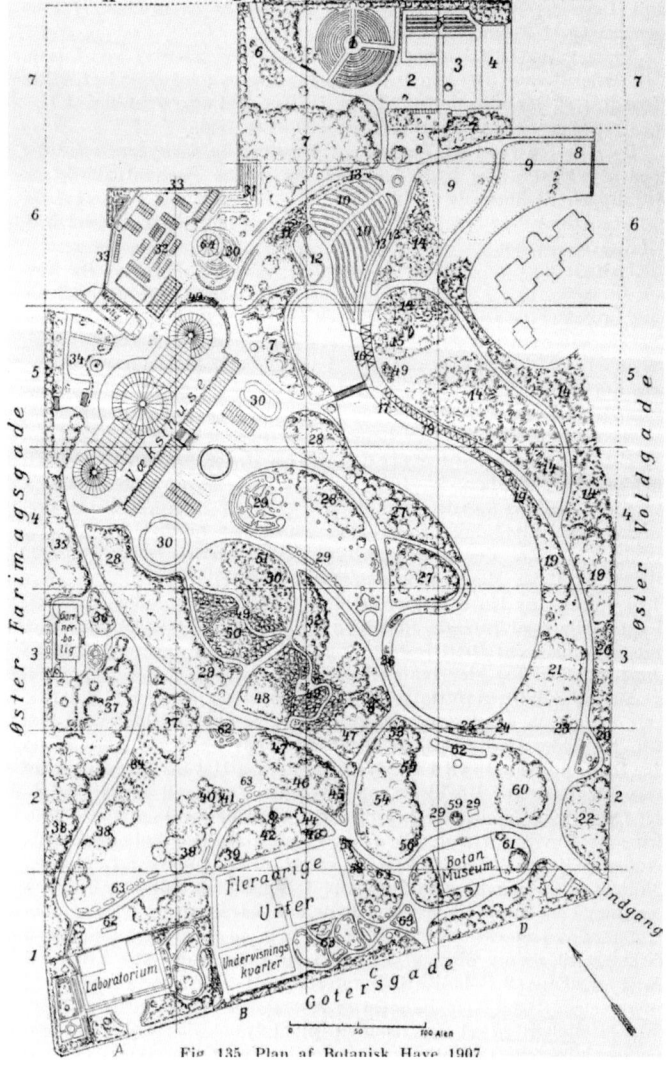

Fig. 135. Plan af Botanisk Have 1907

A New Home

In 1870, the Garden moved to its current
setting, designed by landscape gardener
Henrik August Flindt with Head Gardener
Tyge Rothe, who was involved in the
planning of the spectacular complex of
glasshouses. These were inspired by Sir
Joseph Paxton's Crystal Palace of 1851
and paid for by the Danish industrialist
and founder of the Carlsberg brewing
empire, J C Jacobsen. Jacobsen also
funded the installation of classical
statues around the grounds.

Glasshouses

There are an incredible twenty-seven
glasshouses covering 3,000 square
metres, many of them the historic ones
funded by Jacobsen in 1874. The stunning
glass and iron Palm House was designed
by Peter Christian Bønecke in the same
year on the model of the Palm House at
Kew. 50 metres long and 16 metres tall,
at its centre is a circular section that
doubles in height and looks not dissimilar
to a wedding cake. Inside is a tall, narrow,
cast-iron spiral staircase that takes you

up among the tops of the palms, where the humidity can catch your breath. A massive array of tropical and subtropical palms vie with each other to touch the roof, while below them is an excellent collection of ancient and endangered cycads. The Palm House is also home to the Butterfly House, open during the summer, where the butterflies fill the air: species here include orchids and vanilla plants.

By contrast, you'll need a sweater for the Arctic Greenhouse, whose special cooling air-conditioning system recreates an environment suitable for Arctic plants. Erected in 1959, it is as important today as it was then. The changing conditions of the Arctic are an incredibly important area of research, and the Greenhouse provides an opportunity to study polar plants from the tundra climate. There is also an Alpine Greenhouse, built in 1973, with a further cold-temperature greenhouse added in 1980.

ABOVE: The original, narrow cast-iron spiral staircase of 1874 which takes the visitor up 16 metres (52 feet) inside the Palm House (by architect Peter Christian Bønecke).

OPPOSITE: The steps leading up to the Victorian Palm House and the terrace. From the terrace are wonderful views across the *Botanisk Have* and into the city.

Wunderkammer

As an avid collector, Ole Worm's interests extended beyond the botanical, and he assembled an extraordinary 'cabinet of curiosities' that included everything from ancient scientific instruments to giant animal skulls. Significantly, he established that a purported unicorn horn in his collection was in fact a narwhal tusk. After his death from the plague, Worm's collection was bought by Frederick III of Denmark, who built a museum for its display.

The Ramparts

The Rock Garden is built on the most elevated section of the Garden, an area that once formed the ramparts of the old fortifications of Copenhagen, and gives outstanding views of the city. The lake beneath it, which often freezes over in winter, was once the city moat.

Special Plants

There are a number of monumental trees, including one conifer, a Taxodium, which dates back to 1806. It was moved from an earlier site of the Garden, and it is the oldest tree here. There are also extensive collections of flora native to the Faroe Islands and Greenland, including the rare red-leaved Faroese dandelion, *Taraxacum rubifolium Rasmussen*. The Garden was gifted an *Amorphophallus titanum* (the Titan arum or corpse flower) by the Botanical Garden of Bonn in 2003. Having first bloomed in 2012, it now does so every other year. Blooming is very rare in the wild, so this is a wonderful opportunity to see the fluorescence.

ABOVE: *Botanisk Haves Butik*, the Botanical Garden shop, is located at the entrance to the Garden near Nørreport Station.

Scientific Research

The Garden is home to a Herbarium with the largest collection of dried plants in Denmark – over two million specimens. There are also collections of economic botany, a seed bank and four gene banks, one of which is devoted to the native species of Denmark. It is the only gene bank for wild Danish flowers anywhere. Now fully integrated into the Natural History Museum at the University, high on the Garden's list of priorities is further study towards the conservation of plants and fungi from both Denmark and abroad. The Museum was established in 2004 from a merger of the Zoological Museum, the Geological Museum, the Botanical Museum and the Botanical Garden. A brand new building to house it within the Garden site is scheduled to open in 2025.

ABOVE: The Rock Garden at *Botanisk Have* is spectacular – spot the purple spikes of the dactylorhizas.

Oxford Botanic Garden
England (1621)

The oldest botanic garden in Britain dates from 1621, and occupies a site on the edge of Oxford that was formerly a Jewish cemetery. Founded as a 'Garden for Physical Simples' to teach Oxford University medical students, but also to serve as a horticultural collection, it took some ten years to plant, and was funded by Henry Danvers, 1st Earl of Danby, who commissioned its wonderful (and thankfully still existing) gateways. These were designed by Nicholas Stone, master-mason to two kings: James I and Charles I. The main gate, known as the Danby Gateway, dates from 1633, and in the centre of its pediment is a portrait bust of Danvers, with statues of Charles I and II on either side.

The first horticulturist to take charge of the Garden, John Tradescant the Elder, died within a year of taking the post, and it was not until the appointment, in 1642, of its first Superintendent, Jacob Bobart the Elder, that the garden developed. Bobart had another occupation – he owned a pub on Oxford High Street called the Greyhound. There are some wonderful stories about him, which may or may not be true. One is that he was usually accompanied by his pet goat, and another that on special occasions he would attach pieces of silver to his long beard. He made a scientific list of all the plants and trees growing in the Garden, '*Catalogus plantarum horti medici Oxoniensis*', published in 1648. 1,369 plants were listed, classified by place of origin, and many of these species are still flourishing on the site. The catalogue reveals that both native and rare exotics were being grown, including a plant sent from America, the Virginian spiderwort (*Tradescantia virginiana*). It was named after the Tradescants (father and son), who were naturalists, plant hunters and explorers.

Bobart also planted the yew (*Taxus baccata*) which is now the oldest specimen in the garden – the sole survivor of what was once an avenue of them. His son, Bobart the Younger, also became Superintendent of the Garden, and his herbaria became the founding collection of the 'Oxford University Herbaria', which today contains over a million plant specimens.

Layout of the Garden

The Walled Garden's layout – a square, divided into four with a well in the centre – more or less follows the plan of 1850; there is an extra piece of land that runs to the river (the Lower Garden).

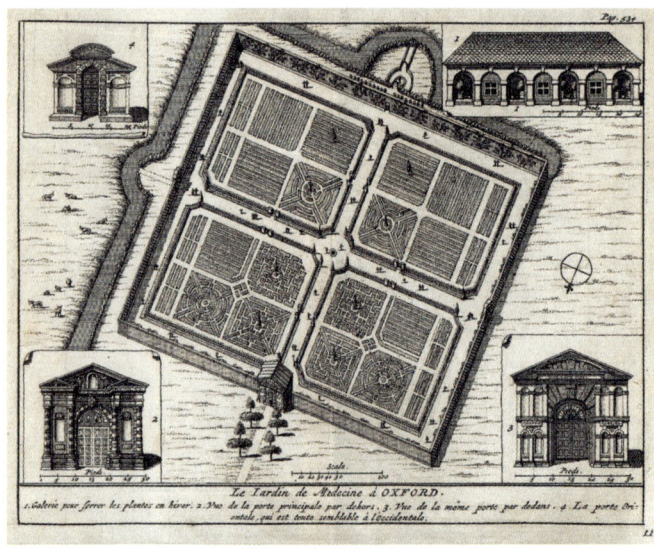

Le Jardin de Medecine à OXFORD.
1. Galerie pour serrer les plantes en hiver. 2. Vue de la porte principale par dehors. 3. Vue de la même porte par dedans. 4. La porte Orientale, qui est toute semblable à l'Occidentale.

The Grade I-listed garden outside the Oxford Botanic Garden, which belongs to Magdalen College (seen on the opposite side of the road). It was designed in 1953 by Sylvia Crowe, with hedges of box and yew filled with roses.

OPPOSITE: A view of the *'Hortus Botanicus, The Phisick Garden in Oxon'* by David Loggan, in the *'Oxonia Illustrata,'* published in 1675.

Within the Walled Garden are the taxonomic beds, the medicinal plant collection, and the Bobart collection, with other areas dedicated to plants from around the globe. In the Lower Garden is a bog garden, as well as a new and inspired Gin Border – rather fitting, as Bobart was a publican. In collaboration with an artisanal distillery, it produces its own Physic Gin, made with 25 botanicals inspired by Bobart's plant list. The sustainable Merton Borders, designed by Professor James Hitchmough and sown from seed, are a highlight of the Lower Garden. Their stunning perennials and grasses were selected to be able to withstand drought conditions, and come from three regions of the world: the Central to Southern Great Plains of America; East South Africa; and Southern Europe across Asia to Siberia. This area needs very little maintenance, and no watering; it is left to overwinter as a garden of architectural skeletons and is cut back in springtime.

Glasshouses

In the 1600s there were citrus fruits growing in the Garden: lemons, limes and citrons which must have been brought inside at the first hint of frost. In 1733 two conservatories were designed by William Townsend (who worked on the Temple of Echo in the garden at Rousham House in Oxfordshire). These enabled the cultivation of 'about 900 exotics in pots', including many plants from the New World: coffee, tea, cotton, sugarcane and pineapples. The first pineapples took some time to grow here, but were successful by 1749. In 1850, the hothouses were replaced by a Water Lily House: water lilies were all the rage at the time, and head gardeners on estates up and down the country were cultivating the newly introduced *Victoria regia* (now *Victoria amazonica*). The Garden currently has seven glasshouses, featuring plants from all over the globe: the Water Lily House survives, though today smaller varieties of lily are propagated – *Victoria cruziana*, and also *Nymphaea x daubenyana* – a hybrid first grown in the Garden in 1874 and named in honour of Professor Charles Daubeny, Keeper of the Garden from 1834 to 1867.

RIGHT: The Water Lily House, built in 1851 especially to cultivate the giant waterlily *Victoria amazonica*. *Victoria cruziana* and *Oryza sativa* are now grown here.

ABOVE: A view from outside the Walled Garden with late 15th-century Magdalen Tower beyond.

OPPOSITE: A late summer border in the Garden with *Echinacea* blooming.

Inspiring scientists and authors – the Garden and literature

Over the centuries, the Garden has attracted many famous visitors, including the diarists John Evelyn and Samuel Pepys, and it features in a number of literary works. In Evelyn Waugh's *Brideshead Revisited*, Lord Sebastian Flyte visits it with Charles Ryder 'to see the ivy' soon after they first meet. Lewis Carroll, who was a professor at Oxford University, was known to spend time in the Garden, and in *Alice's Adventures in Wonderland* the Water Lily House can be seen in the background of the illustration of 'The Queen of Hearts' Croquet-Ground.'

J R R Tolkien, another author who taught at Oxford, was often found in the gardens, sitting and smoking under his favourite tree – an ornamental black pine, said to be his inspiration for the Ent creatures in *The Lord of the Rings*. It survived until 2014.

Will and Lyra's bench from Philip Pullman's *His Dark Materials* can be seen in the Lower Garden. In 2019 a sculpture by Julian Warren, featuring 'Dæmons' from the book, was placed next to it.

A final note

As you leave the Botanic Garden, notice the Grade I-listed garden outside the walls, which belongs to Magdalen College. It was designed in 1953 by landscape architect Sylvia Crowe, and its winding hedges of box and yew are filled with roses. Crowe's commission commemorated the development of penicillin at Oxford University, and its layout echoes that of the Botanic Garden.

Jardin des Plantes
Paris, France (1635)

The *Jardin des Plantes* is laid out as a stunning formal axial garden, in the French manner. It is full of surprises, including a labyrinth, a zoo, museums and stunning Art Deco glasshouses. Located on the left bank of the Seine, in the 5th *arrondissement* of Paris, it covers 28 hectares (69.1 acres) and contains around 10,000 species and varieties of plants, 2,000 trees, and 2,500 shrubs: its grounds have 80,000 seasonal plants, and burst with colour. This is a large site with eleven gardens, so devoting a sunny day to it with a picnic and a map is the best way to discover its many pleasures.

History

The idea for the *Jardin des Plantes* originated with King Louis XIII, who instructed his physician Jean Hérouard to create a medicinal garden for him. Hérouard had seen the first-ever botanic garden in France, opened in the southern city of Montpellier on the orders of Henri IV in 1593. It was attached to a medical school, taught medicinal plant knowledge, and became the prototype for all subsequent French botanic gardens. Hérouard wanted to create something

BELOW: 1636 plan of Paris's 'Jardin du Roy for the cultivation of medicinal plants' by Guy de La Brosse and Frédéric Scalberge.

ABOVE: Behind the green lawns or the *boulingrins* is the Grand Gallery of Evolution which holds the historical zoology collections. It originally opened in 1889.

similar in Paris, and in 1626, Guy de La Brosse was made the new Garden's intendant. He quickly started to acquire seeds and plants from around the world, and after Hérouard's death, it was left to him to present the King with a plan for the garden in April 1634. Staff were brought from Montpellier to help develop it – a controversial move, as the south of the country was politically at odds with the north at the time.

The establishment of the Garden brought plant hunting and collecting under the control of the King, who instructed the French navy to amass exotic specimens from around the globe. In 1636, La Brosse published his *Description du Jardin royal des Plantes médicinales*, listing 1,800 species and varieties then being cultivated in the Garden.

It opened officially to the public in 1640, offering free classes teaching botany, chemistry and anatomy to doctors and apothecaries. To mark the occasion, La Brosse published a pamphlet describing its creation, and comparing it with other important botanic gardens: those of Padua, Pisa, Leiden, and, of course, Montpellier.

Serres chaudes

There are four impressive *serres chaudes*, or greenhouses, constructed from iron and glass. The site's first greenhouse was made around 1713 to house and care for a coffee plant that had been sent to Louis XIV. Two more were built between 1834 and 1836 by the architect Rohault de Fleury on the same site as the 18th century ones they replaced. The most stunning of the greenhouses is the Rainforest Greenhouse, formally known

ABOVE: Interior of one of the Great Glasshouses. The very first glasshouse was built here in about 1713 to house a coffee plant.

OPPOSITE: Rows of Iceland poppies (*Papaver nudicaule*) in the foreground of flower beds of the long parterre.

as '*Jardin d'hiver*' or Winter Garden, designed by René Berger in the Art Deco style, and completed in 1937. It is hot, humid and noisy when you enter, with a fifteen-metre-high waterfall splashing down into a pool surrounded by plants from the tropical rainforests: coffee, vanilla, bamboos and wild banana trees. All the plants date from after 1945, with one exception, the bibby-tree (*Sabal bermudana*), a large palm. Every other plant perished in frosts after the heating was turned off for economic reasons.

The Ménagerie – the zoo of the *Jardin des Plantes*

The *Jardin des Plantes* contains fauna as well as flora. During the early 18th century, the famous naturalist Georges-Louis Leclerc, Comte de Buffon, apothecarist to the King and the director of the Garden (then called *Jardin du Roi*), introduced animals and birds, creating a zoological section. This collection was expanded at the time of the French Revolution, when the palace of Versailles was abandoned, and the survivors of the King's menagerie needed to be relocated. Incredibly, prior to a municipal ruling of 1793, wild animals had been exhibited on the streets of Paris: after this ruling, these animals were ordered to be relocated in the Ménagerie – which opened to the public in 1794, and is one of the oldest zoos in the world. Take a look at its buildings too, as many of them are listed, such as the 1926 Art Deco *Vivarium*, and the Big Cat House, built in 1937 by René Berger – who also designed the *Grande Serre* and the *Grande Galerie de l'Évolution* (Gallery of Evolution).

The Flora

The *Jardin des Plantes'* first garden had fewer than 2,000 species, but today there are over 10,000. The site retains its overall formal design, with a classical avenue of London plane trees (*Allée Buffon*), and the parterres and flowerbeds of the central garden (known as the Grande Perspective) which covers 3 hectares (7.4 acres). Incorporated into the design are the Alpine, Useful Plants, and the Botanical School.

The Alpine Garden (*Jardin Alpin*), originally created in 1640, is a magnificent collection of more than 2,000 kinds of mountain plants. One of the oldest specimens in the garden can be found here: a pistachio tree, planted around 1700, that was used by botanist Sébastien Vaillant in 1718 to prove that plant species have sexual characteristics. The Alpine Garden was enlarged and developed by the Comte de Buffon during the 18th century, though it took on its present shape in the early 1930s. It has a lower-lying, more rugged terrain than the rest of the garden, with a ravine and scenically placed boulders. Its plants come from diverse environments: high, medium and low altitude regions of France and around the world.

The Garden of Useful Plants (*Jardin de Plantes Ressources*) is where, as the name suggests, plants with functional uses are displayed. There are food crops, vegetables, and plants from which pigments can be extracted to use as dyes – like dyer's madder (*rubia tinctorum*), which produces red – and textiles such as linen, hemp and sisal.

L'École de botanique (Botanical School) is still situated in the area where botany was first taught in the Garden. Classification has been an ongoing objective here, and the zone has been reorganized six times since 1635, each time taking into account the latest scientific classification system.

ABOVE: The Alpine Garden, redeveloped in the 1930s, has over 2,000 species of alpine plants from all over the world.

Hortus Botanicus Amsterdam
The Netherlands (1638)

Like many of its early contemporaries, the *Hortus* – which is to be found in Amsterdam's green *Plantage* district – was originally established for the education of doctors and pharmacists. However, its collection soon expanded from the purely medicinal, not least through an influx of exotic plants and seeds supplied by the Dutch East India Company (*Vereenigde Oostindische Compagnie*; VOC).

The History

The *Hortus* was established by the city council in 1638 and Johannes Snippendaal became its prefect in 1646. He undertook a comprehensive cataloguing of the collections, documenting a total of 796 – mostly medicinal – species. However, its remit soon expanded to include specimens of potential commercial interest too.

In the early 18th century, for instance, its greenhouses were used to cultivate propagated coffee seeds collected by VOC traders. Plants descended from these seeds were then taken back to be replanted in South America and proved a mainstay of the burgeoning international coffee trade. The *Hortus* enjoyed similar success with oil palms, growing them from a small number of Mauritian specimens that were then replanted in Southeast Asia.

Architectural innovations included a hexagonal pavilion added in the late 1600s, with an entrance gate coming in the early 1700s. An Orangery was added in 1875. The Garden became the focus of international attention after the Dutch botanist and geneticist Professor Hugo de Vries took over as director in 1885. De Vries gained fame for his 1889 book, *Intracellular Pangenesis*, which drew on the ideas of Charles Darwin and proposed the existence of something de Vries called *pangenes*, a term later shortened simply to genes.

The *Hortus* produces its own honey in beehives to be found in a quiet corner of the garden. They produce just two hundred jars a year, available only in the *Hortus* shop.

The Snippendaal Garden

In 2007 the *Hortus* began a project to recreate the early medicinal collection (the *Hortus Medicus*, as it was known) using Snippendaal's catalogue, of which only two copies survive. It proved a tricky job, as the garden staff had to decipher plant names written using the Pre-Linnaean system. With no records of the original layout either, it was decided to go for a contemporary design but one which evokes aspects of 17th-century garden style. Alongside the predominantly medicinal species are a number of other rare and ornamental plants.

The Semicircle

This part of the *Hortus* was laid out in 1682 as a circular bed of flowering plants but was redesigned along its current lines the following year as a semicircle.

RIGHT: The large glasshouse that overlooks the canal was designed in 1993 by studio ZJA.

Made up of ever-decreasing half-moons of box (*Buxus*) hedge, it is described in the Garden itself as 'a basal arc and three wedges.' This is a systematic garden, in which each 'wedge' – which in summertime is full of flowers and colour – holds plants belonging to a specific classification. The nearer plants are to each other, the more closely related they are. It is both the first and only systematic garden in the Netherlands that categorizes according to the Angiosperm Phylogeny Group (APG) system, based on 'molecular systematics' – that is to say, the similarities between genetic material.

BELOW: The recreated Snippendaal Garden has the same plant species that were planted in the garden in 1646.

The Victorian Palm House

The glass-domed Palm House, commissioned by de Vries and designed in 1911 by architect Johann Melchior van der Mey, boasts a wonderful spiral staircase to take you up to a high walk. It is said that de Vries himself planted both its cinnamon tree (*Cinnamomum burmannii*) and the two ficuses (*F. macrophylla* and *F. lyrata*). In the colder months, lots of the Garden's potted plants are brought inside to await the returning sunshine.

Victoria amazonica

Here in Amsterdam, this aquatic perennial, the giant waterlily, can be found in an outside pond rather than in the glasshouse. The *Victoria amazonica* does flower but is rarely witnessed as it

ABOVE: By the mid-17th century, the collection in the *Hortus Botanicus* included rare and exotic plants, brought from around the world by the Dutch East India Company (VOC).

A Garden for Doctors and Apothecaries

The botanical garden in Amsterdam was originally founded to help find a cure for the rampant bubonic plague, which swept through the Low Countries from the 14th century to the 17th century. When the 'black death' arrived again in the 1600s, the large number of mortalities from the pandemic caused the economy to be rocked. The city decided, in 1638, to found the *Hortus Medicus* as a place to grow medicinal herbs and plants. The *Hortus Medicus* was later renamed the *Hortus Botanicus*.

does so for a very short time, only two nights a year. In the wild, the flower opens in the night and gives off a scent (said to smell like pineapple and butterscotch) that attracts its pollinator – a beetle. The flower then closes in the morning, trapping the beetle until it opens again on the second night to release the now pollen-covered creature to fly off and pollinate. Once pollinated, the flower changes colour from white to pink.

Three-Climate Greenhouse

In contrast to most of the rest of the Garden's buildings, the Three-Climate Greenhouse (designed in 1993 by Zwarts & Jansma Architects, with tropical, subtropical and desert zones) is an exemplar of the contemporary. Perhaps

counterintuitively, it is its desert zone that is its coolest part. In the subtropical section is a collection of South African plants, reflecting a time when the country was colonized by the Dutch. The VOC created the first European garden in what was then the Cape Colony, planting on the north slope of Table Mountain in 1652 to supply their ships with fresh fruit and vegetables. VOC ships helped introduce the *Agapanthus* (or African lily) to the Netherlands, and similarly the scented geranium (*Pelargonium*), *Clivia* and *Gerbera*. There is also a small butterfly greenhouse, where the butterflies can be seen among an array of tropical plants including coffee, cacao, tea, rice and sugarcane.

ABOVE: The canal and bridge frame this view of the newest glasshouse and meadow planting of the *Hortus Botanicus*.

OPPOSITE: Various cacti. Some of the other plants in the glasshouse are quite unique, such as a 2,000-year-old agave cactus and a 300-year-old Eastern Cape giant cycad (*Encephalartos altensteinii*).

Royal Botanic Garden Edinburgh
Scotland (1670)

The Royal Botanic Garden Edinburgh (RBGE), the UK's second-oldest botanic garden after Oxford, started life in 1670 as a small physic garden. Originally occupying a small site next to the Palace of Holyroodhouse, it relocated to its current location at Inverleith in the north of the city in 1820. There are also three sister gardens across the country, at Benmore, Logan and Dawyck, each with its own climate and specializing in plants and trees from different continents. The main Edinburgh site consists of 28.3 hectares (70 acres) of landscaped gardens with a Chinese Hillside, woodland, arboretum, Rock Garden, waterfalls and glasshouses.

History
The original Garden was the creation of two local doctors, Sir Robert Sibbald and Sir Andrew Balfour. Funded by the University of Edinburgh, within five years it had grown so much that it moved to a new site next to Trinity Hospital, on a plot where Waverley Station now stands. By 1684 it was reported to contain some 2,000 non-indigenous plant species; but, in 1689, disaster struck when Edinburgh Castle was besieged and a dam at the east end of the nearby Nor' Loch broke, flooding the Garden and destroying all but a few of its hardiest specimens.

The renewed Garden received its Royal Warrant in 1699 and expanded

RIGHT: The wonderful Victorian Temperate Palm House, designed by Robert Matheson, opened to the public in the Garden in 1858.

BELOW: Interior of the plants and people glasshouse where the giant *Victoria amazonica* lilies are grown from seed sown each year.

significantly under Dr John Hope (Regius Keeper 1761–86). In 1763, he oversaw the transfer to a new 2-hectare (5-acre) site on the west side of Leith Walk. There was still further expansion under Daniel Rutherford (Regius Keeper 1786–1819), while during the First World War, one of his successors, Isaac Bayley Balfour, pioneered the use of sphagnum moss to dress wounds.

The Herbarium

The RBGE's Herbarium is older than even that of Kew and contains over 2 million specimens from all over the world. It boasts one of the finest collections of preserved *Meconopsis*. The blue Himalayan poppy has been grown at the RBGE since 1867, when seeds were sent from Scottish botanist-surgeons working for the British East India Company.

All four of the RBGE gardens have collections that flower in late spring. The new Herbarium and Library building, opened in 1964, contains a collection of drawings and watercolours from India by indigenous painters (many of whose names have been lost) known as Company Artists.

George Forrest and the Rhododendrons

Scottish botanist George Forrest (1873–1932) worked for a short time in the Herbarium before undertaking seven plant-hunting missions to Yunnan in southwest China. Incredibly prolific, he brought back over 31,000 herbarium specimens in total. His work is reflected in the collection of rhododendrons across all

four of Scotland's royal botanic gardens, which cultivate around half the known species. Forrest brought back over 300 varieties from China, where previously only 150 had been known. Since the 1970s, the RBGE has taken part in expeditions to Indonesia to investigate tropical Vireya rhododendrons, and houses the world's largest collection of cultivated specimens.

Another highlight of the Garden is the Chinese Hillside, a rugged landscape recreating the terrain of southwest China and filled with over 1,500 plants from the region. Opened in 1997, its waterfall snakes through coniferous woodland and a Rhododendron forest, as well as alpines and meadow plants, to an *Iris* and *Primula*-surrounded pond overlooked by a Ting (pavilion) for rest and contemplation.

LEFT: Inside the Victorian Tropical Palm House at RBGE. Some plants are kept in pots: until the 1890s, even the largest palms would be housed in this way.

BELOW: The Chinese Hillside in a blaze of fiery colour in the autumn. Planted in 1997, this garden has the largest collection of Chinese plants in the western world.

The Rock Garden

When the Rock Garden was built in 1871 by the then curator, James McNab, it was something of a pioneer project. Designed for true alpines (those that have evolved to survive at high altitudes) – a new idea at the time – it consisted of small compartments, each planted with single, clearly labelled specimens. The Rock Garden that exists today, home to some 5,000 species, was completed in 1914 with more natural plantings. In 1933 a scree bed (where grit and rubble are dug into the soil) was developed based on George Forrest's experiences in China; this was particularly suited to alpines.

RIGHT: The Titan arum (*Amorphophallus titanum*), also called the 'corpse flower' as it smells like death, in bloom at RBGE.

ABOVE: The John Hope Visitor Centre was designed to be sustainable. Its glass walls allow views into the Biodiversity Garden.

OPPOSITE: The stream and waterfall that run through the Chinese Hillside and finally collect in a large pond.

Glasshouses

The garden's first glasshouse was built in 1713 on the designs of George Preston. In 1834, an 18-metre-wide, 8-metre-high octagonal Tropical Palm House complete with eighteen stone pillars to provide shade was built at a cost of over £1,500. At the time, its conical roof, rising to 14.3 metres, was the tallest in Europe but it was soon deemed too small, so a Temperate Palm House designed by Robert Matheson was added in 1858. In 1967, the complex was expanded when Princess Margaret opened the Front Range glasshouses. Today, the Tropical Palm House houses a *Sabal bermudana* specimen – the oldest plant in the RBGE collection, believed to date from around 1790. Additionally, there is an Alpine House which uses a fan to mimic windy mountain conditions.

The Incredible stinking plant

The Tropical Glasshouse has been home since 2003 to a Titan arum (*Amorphophallus titanum*), a variety first recorded by Italian botanist Odoardo Beccari in the jungles of western Sumatra in 1878. A giant standing 3 metres (nearly 10 feet) high, it is often called the world's largest flower, but this is a misnomer as it produces an inflorescence (male and female flowers). Its unfortunate nickname is the 'corpse flower' because of the stomach-turning odour of decay it produces to attract its pollinator, the carrion beetle. It flowered for the first time at RBGE in 2015.

Conservation and Education

The garden's scientists and botanists work with an international network to counter the threat of species loss. The RBGE has a particular interest in conifers, the oldest tree type in existence, and in 1991 hosted the creation of the International Conifer Conservation Programme. It is possible to study here for an Edinburgh University MSc in Biodiversity and Taxonomy of Plants, and, like Kew and Wisley, the RBGE runs training schemes for gardeners.

Chelsea Physic Garden
London, England (1673)

Hidden behind high walls in Chelsea is London's oldest botanic garden, one that was created for the advancement of medicine and the education of its practitioners. Covering 1.4 hectares (3.45 acres) of prime London real estate, it aims to be 'a physic garden for the future', highlighting the ongoing importance of plant science to our understanding of climate change and biodiversity.

The History
Back in 1673, when the Worshipful Society of Apothecaries chose the site of their garden, there was access straight onto the River Thames (later impeded by construction of the Embankment). Its purpose was to train apothecaries to recognize the herbs used in medical treatments. The location was perfect. It had previously hosted a market garden, so the soil was known to be fertile. Meanwhile, the river facilitated the transportation of plant specimens and provided mooring for the Apothecaries' barge, which was used for plant-hunting expeditions.

The wall around the Garden was built around 1676–7, providing it with security and creating a microclimate. Its warming south-facing aspect and rich soil supports plants that struggle elsewhere in the city. In 1685, the first Cedars of Lebanon (*Cedrus libani*) to grow successfully in England were planted here, and rare rose species like the *Rosa chinensis* (Bengal Beauty) from China, with its pretty single scarlet flowers, can bloom all year round.

Benefactors, Curators and Directors
In 1712, Sir Hans Sloane (1660–1753) purchased the Manor of Chelsea from Charles Cheyne. A physician, botanist and collector, Sloane's treasures

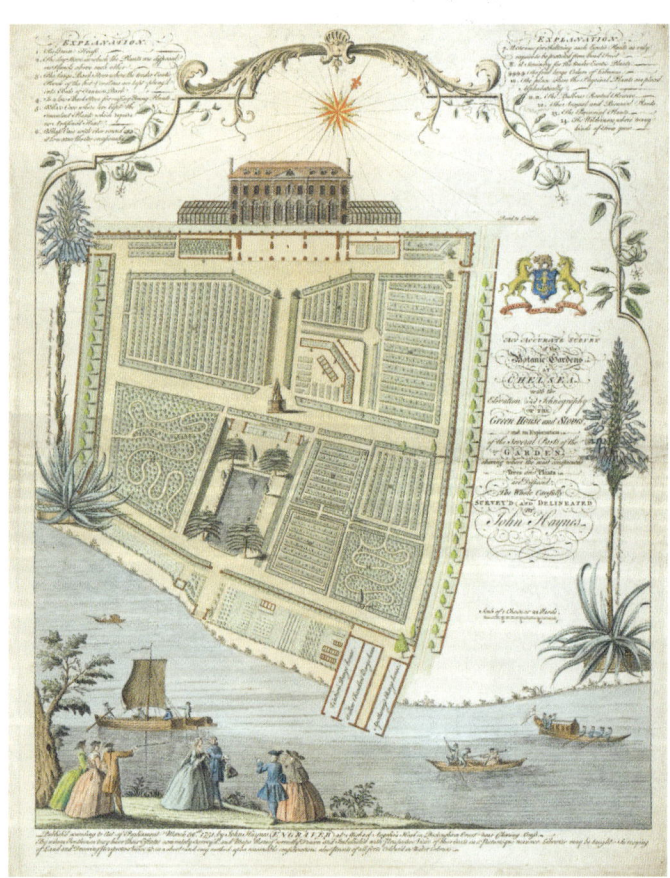

ABOVE: The Physic Garden, Chelsea: a plan view. Engraving by John Haynes, 1751.

OPPOSITE: The Pond Rockery, a Grade II-listed structure, is said to be the oldest in Europe.

provided the foundation for the British Museum. As physician to the Governor of Jamaica, the Duke of Albemarle, he witnessed slaves working on the sugar plantations and was also introduced to cocoa. Finding it too bitter, he mixed it with milk and sugar, which apothecaries in England then sold as a medicine.

Sloane leased the Garden in Chelsea to the Worshipful Society of Apothecaries of London for just £5 per annum in perpetuity. All he asked in return was that it supply the Royal Society, of which he was a principal, with 50 herbarium samples per year, up to a total of 2,000 plants. Without his generosity, the garden would not exist today in one of London's most expensive neighbourhoods.

Sloane appointed Philip Miller as Head Gardener in 1722 and under him the Garden became famous across Europe for its rare and exotic species (not all medicinal). Visitors included the Swedish botanist Carl Linnaeus. Over almost fifty years, Miller cultivated all kinds of exotic fruits, including pawpaws, melons and pineapples. The latter – incredibly rare in Britain at this time – were grown in hot beds or pits made from fermenting oak bark.

William Forsyth (an ancestor of the late Sir Bruce Forsyth) joined the Garden in 1771 and created the Grade II-listed Pond Rockery that still stands today. Built in 1773, it contains stones from the Tower of London as well as Icelandic basalt given by Joseph Banks. Rare, endangered and unusual species from southern Europe and North Africa thrive here.

In 1846, Scottish plant-hunter Robert Fortune (famous for smuggling tea plants from China to India) was appointed Director, and oversaw major changes, like the creation of Fortune's Tank which still provides a habitat for frogs and toads that keep down the slug population.

ABOVE: The Pond Rockery, dating from 1773. It has pieces of stone from the Tower of London, basalt brought back by Joseph Banks, and even clam shells from the HMS *Endeavour*'s trip to Tahiti.

Hot-Housing
In the 1680s Chelsea erected its first glasshouse, the first heated one in the country. In 1685, gardener and diarist John Evelyn wrote that he had seen a Cinchona tree (*Cinchona ledgeriana*) there. This was the source of quinine, a drug used in the 17th century to combat malaria. Sir Hans Sloane incorporated it into an ointment to treat eye problems.

Economic Plants

The Garden has grown many economic plants including rubber, cocoa and coffee, acknowledging the overlap between horticultural innovation and Britain's exploitation of colonial plant resources. This complex and uncomfortable relationship was explored in the Glasshouse Restoration Project, which overhauled the Garden's Foster & Pearson glasshouses, built in 1902, in time for the Garden's 350th anniversary in 2023.

Index Seminum

From 1682, the Garden became known around the world as it initiated a seed exchange with other botanical gardens, using a seed catalogue called *Index Seminum*. The idea sprung from curator John Watts, who invited Dr Paul Hermann of the Leiden University Botanic Garden to Chelsea. Hermann returned to Leiden laden with plants and seeds, beginning an exchange that continues today and now encompasses 368 gardens and universities in 37 countries. Despite being the most important centre for plant exchange in the world in the 18th century, Chelsea's first *Index Seminum*, as we would recognize it now, was only published in 1901.

Medicinal Garden

Planted in the area where apothecaries grew medicinal plants in the 17th and 18th

LEFT: *Agapanthus* growing outside the Victorian glasshouse (before restoration).

centuries, this Garden was the creation of 19th-century curator Thomas Moore. It is divided into thematic sections including the World Medicine Collection (comprising specimens used by healers over the last 5,000 years) and the Dioscorides Bed (which contains plants in the *Materia Medica* of the Greek pharmacologist, physician and botanist, Dioscorides).

The Hot Wall

Planted along the warmest wall of the garden, which runs beside Swan Walk, is a collection of rare and endangered plants. Among them are species from Crete, Madeira and the Canary Islands, including the biennial *Echium pininana*, whose towering spire is 2.4m (8ft) tall, and the rare *Echium wildpretii* from the top of the volcanic Mount Teide in Tenerife. Many of these species thrive today thanks to the Garden's unique microclimate.

Chelsea is also home to the UK's largest fruiting olive tree – which in 1976 yielded a record crop of 3.17kg (nearly 7lb) – and the largest outdoor fruiting grapefruit.

The Dicotyledon Order Beds

The name 'Dicotyledon' refers to a group of flowering plants which have two seed leaves at germination, which this part of the garden focuses on. The Order Beds were laid out in 1902, and today there are over 800 plants arranged according to their families. This classification will change in the not-too-distant future as scientists and botanists begin to classify plants according to their DNA sequence.

Real Jardín Botánico
Madrid, Spain (1755)

On a wide street in Madrid are some of the most important cultural institutions of the city, including the Botanic Garden. This didn't happen by happy accident, but due to a monarch living in the time of the Enlightenment in Spain. In 1774, King Charles III decided he wanted all the educational centres within easy access on one thoroughfare, which resulted in the relocation of the Garden to where it resides today, next to the famous Prado Museum.

Divided into three major terraces and with numerous glasshouses, the Garden contains some 30,000 plants and flowers and 1,500 trees. In 1942 it was declared a Historic Garden and is registered in the Catalogue of Assets of Cultural Interest of the Spanish Cultural Heritage. Its research work is currently focused on studying the diversity of plants, fungi and ecosystems with a view towards better conservation practices.

OPPOSITE: On the terrace of the *Plano de la Flor* is the Villanueva Pavilion, now used as an exhibition space, and in front of the pavilion, in the centre of the round pond, is a bust of Carl Linnaeus.

BELOW: Tulip beds in springtime in Madrid at the *Real Jardín Botánico*.

The History

The city's first botanic garden, situated at the Orchard of Migas Calientes (modern-day Puerta de Hierro), was established in 1755 on the orders of King Ferdinand VI. Curated by the botanist and surgeon José Quer y Martínez, it contained around 2,000 plants. Just a few years later, Charles III came to the throne and oversaw a grand modernisation of the Spanish capital. His vision included the creation of a science, education and culture hub that came to include not only the Botanic Garden and the Prado but also the National Museum of Natural Sciences (originally known as the Royal Cabinet of Natural History) and the Royal Academy of Fine Arts.

The 'New' Botanic Garden

The new site for the Garden was surrounded by a wrought-iron railing, with its triple terraces laid out by, firstly, the royal architect Francesco Sabatini, and then, from 1780, Juan de Villanueva (who also designed the Prado).

It was soon well stocked with specimens brought back from Crown-sponsored expeditions abroad, particularly to the country's colonial outposts, in search of goods and botanical material of economic value. For the live specimens (as opposed to seeds or dried specimens) that survived the voyage back to Spain, Madrid did not always present a hospitable climate, and many were found to fare better at sister gardens in, for instance, Cádiz, the Canaries and Aranjuez. Nonetheless, Madrid became a focal point for research into the scientific and commercial potential of many newly-introduced species, with gardens in Barcelona and Valencia annually receiving supplies of seeds from the capital.

The Spanish War of Independence proved a catastrophe for the Garden, which was all but abandoned in 1808. Only the efforts of its director, Mariano Lagasca – himself forced into exile for several years – saved it from falling into utter disrepair. A renaissance came under one of his successors, Mariano de la Paz Graells y de la Agüera, in the 1850s. His achievements included the addition of a new greenhouse and the opening of a zoo (later relocated to the *Parque del Buen Retiro*).

The Garden was beset with more problems in the 1880s when it lost part of its footprint to the Ministry of Agriculture and then suffered a devastating cyclone that took down over 500 trees. Come the 20th century, it fared little better, suffering the consequences of civil and world wars. In 1974, a shadow of its former self after decades of neglect, it was closed to the public. But under the inspirational stewardship of Leandro Silva Delgado – himself a pupil of the Brazilian landscape architect, Roberto Burle Marx – the Garden was reborn and triumphantly reopened in 1981.

OPPOSITE: Interior of the Graëlls Greenhouse, which dates from the 19th century. Its collection of tropical plants is a testament to the efforts of former directors to preserve the *Real Jardín Botánico*.

The Three Terraces

The first terrace you come to is adorned with ornamental plants, growing within box-edged beds. These are divided into collections of medicinal and aromatic flora, endemic species and fruit orchards (complete with an insect hotel to maintain the pollinator populations), with a fountain serving as a uniting focal point. The terrace, which also has a rockery, is designed with education as its primary function, so no edibles are harvested here but are instead left to grow and decay observably.

The second terrace, The Terrace of the Botanical School, presents a taxonomic collection of plants, ordered phylogenetically (according to evolutionary relationships), its beds arranged around twelve small fountains.

Meanwhile, the third terrace – the Terrace of the Flowers, designed in the mid-nineteenth century in the romantic English style – contains the Villanueva Pavilion. Designed by Juan de Villanueva and built to serve as a greenhouse in 1781, the Pavilion was declared a historic monument in 1942 and now functions as an exhibition space, with a café and shop. In front of the Pavilion is a large pond overlooked by a statue of Carl Linnaeus, that genius of taxonomy. Above is a slope that takes you to an area of Bonsai trees, where Asian and European species are arranged on plinths that loop around a large pool at the top of the Garden. They were donated in 1996 by the former Spanish Prime Minister, Felipe González, a keen Bonsai hobbyist in his spare time.

The Glasshouses, Herbarium and Germplasm Bank

The Garden has two glasshouses. The older Graëlls Greenhouse dates from the 19th century, while the more modern Three Climates Greenhouse exhibits plants from tropical, temperate and desert environments. The Herbarium, meanwhile, is not visitable in person but is accessible online and holds in excess of a million specimens, many dating back to the era of colonial expeditions. There is also a Germplasm Bank that keeps a collection of 2,500 wild species seeds at low temperatures and low humidity to preserve them in optimum condition in the interests of conservation of genetic diversity.

BELOW: The 'Santiago Castroviejo Bolíbar' greenhouse is named after a former Garden director. It has plants from desert, subtropical and tropical climates.

Palmengarten
Frankfurt, Germany (1763)

The motto of the *Palmarum hortus* is 'Plants, Life, Culture', celebrating the unique character of this Garden, and its importance for the community. A botanical haven located in the centre of Frankfurt, it houses plants from all over the world as a source of education and an oasis of calm within the city. It also provides a stunning setting for regular music performances and concerts.

The Garden's displays and collections now hold more than 13,000 subtropical and tropical species, cultivated in themed gardens and arranged according to plant types in the greenhouses (7,000 sq.m/75,347 sq.ft) and the landscape grounds which stretch over 22 hectares (54 acres).

The History

The first botanic garden in Frankfurt was founded in 1763 by Dr. Johann Christian Senckenberg as a *hortus medicus* for his Senckenberg Foundation. It moved twice before being set up on the present site.

What we know as the *Palmengarten* was originally established after the 1866 Austro-Prussian War, when Duke Adolphe of Nassau was exiled from his estate at Wiesbaden-Biebrich, west of Frankfurt. This prompted him to sell his exotic and tropical plant collection, with its extensive greenhouses and world-famous conservatories. Garden designer and botanist Heinrich Siesmayer (1817–1900) was commissioned by the Duke to liquidate the collection, but Siesmayer could not resist the opportunity of creating a garden himself. He wanted to make a 'Winter Garden' like those popular

in Britain at the time – a glasshouse filled with foreign plants, suitable for concerts, dancing and taking tea.

In May 1868 a committee was formed to help fund the *Biebricher Wintergärten*, and when shares were issued, they proved so popular that the committee was able to buy the Duke's valuable plant collection. The City of Frankfurt gave about 7 hectares (18 acres) of land to them, and, under Siesmayer's supervision, the glasshouses were erected by 1869, and hosted their first flower show in 1870. The 'Palm Garden' is now one of the largest of its kind in Europe, and still functions in the 'Winter Garden' tradition, holding concerts and functions.

Siesmayer was director from 1868 to 1886, and was succeeded by August Siebert (1854–1923), a great horticultural and botanical expert. Siebert was to expand and improve the site over his forty years in charge, building new greenhouses and installing electricity. He was also responsible for the first published companion to the *Palmengarten*, which appeared in 1895.

The War periods

During the First World War, the greenhouses and grounds were used as vegetable plots to supply military hospitals. Although the Garden managed to be maintained during the war, the

RIGHT: The *Palmengarten* allows visitors the chance to see and experience a great diversity of plant species from all parts of the world.

OPPOSITE: The Palm House opened in 1869 as a Winter Garden, complete with a ballroom.

subsequent economic crisis required major financial cutbacks. The outbreak of the Second World War caused the gardens to revert to growing potatoes and cabbage, and when Frankfurt was attacked from the air in 1944, the *Palmengarten* was not spared: the western section of the *Gesellschaftshaus* (literally, 'society house' – an impressive building with a ballroom, opened in 1871), the music pavilion and all the glasshouses were destroyed by bombs and fire.

After the war the occupying American forces used the *Gesellschaftshaus* and park as a place of military recreation. They eventually rebuilt and repaired all structural damage, and in 1953 the Americans returned the gardens to the City of Frankfurt. The Director of the Garden at this time was Fritz Encke (1904–2000): he had previously worked as Frankfurt's Garden Inspector. Under Encke the *Palmengarten* was reopened and went from strength to strength; he also expanded the plant collection and

began participating in seed exchanges with other gardens all over the world. In 1963, one million visitors were counted in one single year alone, reflecting its enormous and growing popularity among Frankfurt's citizens.

Gesellschaftshaus restoration

The restoration of the glasshouse turned out to be a mammoth undertaking: The 35-million-Euro project, based on plans by David Chipperfield, was a massive construction project. The façade, which had been covered up in 1929 in the German 'New Objectivity' style and later changed again, received a complete makeover. The magnificent ballroom, with its historical structures and wall paintings, was restored to its original splendour, and the staircases and technical installations were all modernised. The Society House's grand reopening was celebrated with a gala event in 2012.

The *Tropicarium* and botanical inspiration

The *Tropicarium* at the *Palmengarten* was designed by Hermann Blomeier, and completed in two stages: *Tropicarium North* opened in 1984, and *Tropicarium South* in 1987. In designing the building, Blomeier was inspired by the cross-section of a *Cereus* cactus. The sophisticated structure has computer-controlled, cutting-edge technology, such as its heating system. Hot water pipes are built into the frame, and ten tons of water circulate around the building, warming the air. The plants are nurtured while being watered, as rainwater

tanks under the building are heated automatically, and the exotic plants are gently sprayed with water of the correct temperature. Within the *Tropicarium* there are ten different climatic zones which take the visitor from tropical rainforests with orchids and palms to the desert regions with succulents and cacti.

The 'Visual Axes'

The Garden has introduced a wonderful installation, making it possible to see the site as it was when it was founded, and to compare it with the present view. There are fuchsia-red 'Stelae' (slabs), each with two frames. One displays a

ABOVE: The *Palmengarten*'s *Gesellschaftshaus* steel and glass construction was inspired by the buildings of the Paris World Fair.

RIGHT: *Monstera deliciosa,* one of around 13,000 plant species in the Garden.

historical photograph; the other is empty, 'framing' the view as it is today. Called 'Visual Axes', the installations are an aid to understanding the history of the park, and give an idea of how it has changed over the past 150 years.

Sir Seewoosagur Ramgoolam Botanic Garden
Pamplemousses, Mauritius (1767)

The oldest botanic garden in the Southern Hemisphere is commonly known as the Pamplemousses Botanic Garden, referencing its location in Pamplemousses, about 13km (8 miles) northeast of Port Louis, Mauritius. The area is named after the grapefruit tree, *Pamplemousse* or *pamplemoucier* (*Citrus x paradisi*) which grows in the region, having possibly been introduced from Java by the Dutch. Managed by the SSR Botanic Garden Trust since 1999, the main objectives of the Garden are to promote conservation, education, recreation, culture and history. It is officially known as the Sir Seewoosagur Ramgoolam Botanic Garden, taking its name from the Mauritian doctor who had a distinguished political career as Chief Minister, first Prime Minister, and fifth Governor-General of Mauritius, and was cremated in the Garden.

Notice the white Victorian cast-iron gate and the railings running along the walls – a feature that won first prize at the International Exhibition in London in 1862.

The History
Mauritius was colonized by the Dutch, the French and finally the British, only gaining independence in 1968. During the French period (1715–1810), when Mauritius was known as 'Isle de France', its Governor, Bertrand-François Mahé de La Bourdonnais, had the *Château de Mon Plaisir* built: it stood just where the main gate of the Garden is today. It contained not only his private garden, but also the site that grew cassava (*Manihot esculenta*), which the Governor had brought from Brazil to provide food for the island's slaves.

It is the aptly named Pierre Poivre (his surname means 'pepper') whom we can thank for founding the Botanic Garden. In 1767, this French botanist, while serving as *Intendant* (administrator) for the island, was growing vegetables, fruits and flowers from around the globe in what was then the *Jardin du Roi* (the King's Garden). He was assisted there by Jean-Nicolas Céré (1775–1810), a botanist who would take over its Directorship. Together they worked on a spice collection, cultivating some of the most valuable plants of the period, such as nutmegs (*Myristica fragrans*) and cloves (*Syzygium aromaticum*) from Malacca – species still represented in the large Spice Garden at Pamplemousses today.

During the 18th century, the Garden was maintained as a nursery for acclimatizing crop plants with economic potential, mostly from Europe and the East.

Resurgence
After a subsequent period of decline, the Garden was revived with the arrival of James Duncan as director in 1849. He was responsible for introducing a large collection of palms, including the majestic Royal Palm (*Roystonea regia*). A letter dated 27 March 1853 from Duncan to Sir William Hooker at Kew provides insight

RIGHT: The Lotus Pond full of waterlilies; all of the species here come from China.

into the Garden as it was then. He writes of becoming familiar with its plants and the local language, comments on the state of the Garden and its unpopularity on his arrival, but adds that 'great progress has since been made', and that the governing Council have recently voted for the provision of six 'additional Indians' to work there; this is as well as the group of prisoners who have been assisting him in enclosing the site and installing a gate at the entrance. He says he distributed more than 16,000 plants and trees during the previous year, and expects to exceed that number in the present one. He also assures Hooker that he is working much harder here than in England.

The *Château* and its Trees

The *Château de Mon Plaisir*, built in 1823 on the site of La Bourdonnais' earlier house, is listed as a Protected National Monument of Mauritius. In front of it is a collection of trees planted by dignitaries such as Nelson Mandela, Indira Gandhi, Princess Margaret and François Mitterrand.

The main avenues were built during Céré's time in the late 18th century: he was also responsible for installing several ponds, including the incredible giant one that, since 1927, has been filled with *Victoria amazonica*. There are some 650 varieties of plants to see, among them the famous Baobabs, the *Palmier bouteille* (bottle palm), 85 other kinds of palm trees from around the world, and a collection of medicinal plants. Recently, it has been recorded that various plant species in the Garden are beginning to show phenotypic changes due to alterations in temperature and rainfall. The seasonal pattern of flowering plants has also altered over the years, with tropical species like

ABOVE: The old colonial *Château de Mon Plaisir* has now been restored and is called *Château de Labourdonnais*.

OPPOSITE: The white Victorian cast-iron gate and railings won first prize at the International Exhibition in London in 1862.

Coco de Mer

In 1795, visiting French botanist Jacques Labillardière noted that the rare *coco de mer* palm, *Lodoicea maldivica*, was growing in the Garden. Endemic to the island of Praslin in the Seychelles, it is botanically notable for its 'double' coconut, the largest seed in the world, weighing up to 18kg (39.6lb). The species' male and female are borne on separate trees, with both needed to guarantee reproduction. Those that Labillardière had seen were planted in 1769, not long after the plant's discovery on an expedition to the Seychelles organized by French explorer Marc-Joseph Marion Dufresne.

the Scarlet Flame Bean (*Brownea grandiceps*) blooming earlier each year.

There are animals too, including deer from Java introduced by the Dutch, and Aldabra giant tortoises from the Seychelles, they were brought here in 1875 in an early conservation attempt to protect them from being preyed upon by seabirds and hunted by humans.

ABOVE: The huge trunk of a baobab tree (*Adansonia digitata*). This species dates back to prehistory, and is found all over Africa.

RIGHT: Unlike so many of the gardens in this book, the Sir Seewoosagur Ramgoolam Botanic Garden has no need for glasshouses to grow its *Victoria amazonica* giant waterlilies.

Acharya Jagadish Chandra Bose Indian Botanic Garden
Kolkata, India (1787)

Founded in 1787, the Royal Botanic Garden, Calcutta, was the leading centre for botanical scientific investigation in the British South Asian colonies, and became one of the largest tropical gardens in the world. It changed its name to 'Indian Botanic Garden' following the country's independence, and was renamed again (after plant physiologist, physicist and natural scientist Sir Jagadish Chandra Bose) in 2009. When discussing the Garden's history, we will use its original name, and the anglicised spelling of 'Calcutta' which was officially replaced in 2001 with the Bengali spelling and pronunciation: 'Kolkata'.

Science, Empire and the Natural World

The Garden was established by Lt-Col Robert Kyd, an employee of the East India Company. Kyd was a military man with no scientific training, and his interest in plants was purely as an amateur, sketching his collection in the garden of his home near Calcutta. Nevertheless, he recognised the necessity for a scientific garden in India – a place where plants could be grown and examined for their uses. To have the project taken seriously, and to obtain funding, he needed the East India Company to get involved. In a letter to them dated 1 June 1786, he explained that a botanic garden would 'enable us to outstrip our Rivals in every valuable production which nature has confined to this part of the globe...'

Kyd's efforts were successful, and a 121-hectare (300-acre) site for the Royal Botanic Garden was chosen on the western bank of the Hooghly River, a few kilometres from the centre of Calcutta, and close to Kyd's own house. It came to be regarded as the greatest of all the colonial botanic gardens, remains the largest botanical garden in India, and is still a vital place of study.

During the colonial period, all the botanical research, collection and experimentation of the Botanical Survey of India (BSI) was concentrated here. BSI was established in 1890 with the aim of investigating India's plant resources, and identifying plant species that could be exploited for economic purposes. It now is run by the Indian Ministry of Environment, Forest and Climate Change, and continues to study and conserve native Indian flora.

The East India Company's involvement

The East India Company was prepared to finance the garden because it allowed them to utilize the rich and rare plants grown locally. The original aims were to collect species of commercial value, including spices, teak and – most importantly – tea. The Garden was also ideally positioned to act as a stop-off point where plants heading to and from Europe could become acclimatized before they continued their travels.

OPPOSITE: This Great Banyan tree, which is said to be more than 250 years old, is the world's largest. It occupies a total area of about 2 hectares (almost 5 acres).

In 1770, John Ellis published a 'how to' manual, directed at the 'Captains of Ships, Sea Surgeons, and other curious Persons', and entitled *Directions for bringing over seeds and plants, from the East-Indies and other distant countries, in a state of vegetation*. At this time, plants and seeds were unlikely to survive long sea journeys. When the first fruit trees from Europe arrived in the Calcutta Botanic Garden in June 1787, the majority of them were dead, and none survived in the garden.

Joseph Banks

Back in England, Joseph Banks at Kew Gardens was a great supporter of the Indian garden, which he envisaged as part of a planned network of similar colonial resources – gardens where newly found plants could become acclimatized and then experimented with to work out what they could be used for. Banks was hopeful that they would discover which crops could be grown to boost the economy, and also saw the potential if botanic gardens were to exchange and swap plants and seeds in a more organized way.

The founding father of Indian botany

After Kyd's death in 1793, William Roxburgh (1751–1815) became the second superintendent of the Garden: unlike Kyd, he was a doctor and a botanist. Roxburgh had previously held the post of 'Botanist in the Carnatic': in this South Indian region, he had conducted economic botany experiments, growing plants such

as coffee, pepper, cinnamon and indigo. He had also employed local artists to illustrate plants; they produced beautiful watercolours in what became known as 'Company Style' – a mingling of Indian and European influences. The East India Company needed these illustrations to help them understand what plants were being found and grown in the botanic garden. Roxburgh understood botany's importance in identifying plants to be used in construction, textiles and other areas of commerce. During his tenure the Calcutta Botanic Garden developed into a centre for knowledge, and began specializing in native flora from all the regions of India, becoming a place to which plants could be sent from across the subcontinent to be identified. In 1793, he began gathering Indian plants to start a herbarium, and it became a vast collection and the core of the present-day Central National Herbarium of India. As the Garden made advances in cultivating plants, such as orchids and palm trees, it began to supply European gardens, including Kew.

The collection today

The Garden is a retreat from the noise the pollution of Kolkata: a haven of tranquillity, full of wildlife and birds. The best time to visit it is between September

RIGHT: The Roxburgh Monument at the Acharya Jagadish Chandra Bose Indian Botanic Garden. It commemorates surgeon and botanist William Roxburgh, who was superintendent of the Garden for 20 years from 1793.

ABOVE AND RIGHT: The Great Banyan tree (*Ficus benghalensis*) in Acharya Jagadish Chandra Bose Indian Botanic Garden. It has over 3,500 aerial roots but no main trunk.

and March, when the sun isn't searingly hot. It possesses an incredible 14,122 plant species, including trees, shrubs and climbers, fed by the water from twenty-four interconnected lakes which connect to the River Ganges. The Large Palm House has unusual specimens from around the world, including the Double Coconut Palm *Lodoicea maldivica* (Seychelles), the Cannonball Tree *Couroupita guianensis* (tropical forests of Central and South America), and the very rare Indian Mad Tree, so called because none of its leaves are the same (*Pterygota alata var. irregularis*).

The Great Banyan tree

The Garden's greatest draw is its Great Banyan tree (*Ficus benghalensis*). It is over 250 years old, and survives without its main trunk, which had to be removed in 1925. The Guinness Book of World Records recorded it as the largest tree specimen in the world in 1989.

Parque de Monserrate
Sintra, Portugal (1789)

A botanic garden with a difference, Monserrate is a combination of wild rugged landscape with ruins and waterfalls, formal lawned areas, and cacti and succulent gardens. The Garden sits on the lower slopes of the Sintra Mountain, with valleys that create their own microclimates. Sintra has one of the mildest climates in Europe, so the Garden is frost-free, and can grow tender exotics outside in the wider landscape. At its centre is a stunning palace (*Palácio de Monserrate*), which has a distinctive mixture of different architectural styles. It has been the site of various buildings and gardens for hundreds of years.

The History

The land on which the Garden sits belonged to the *Hospital Real de Todos os Santos* of Lisbon, and in 1540, the hospital's rector, Friar Gaspar Preto, built a hermitage there, dedicated to Our Lady of Monserrate. He envisaged it also having an area where produce for the hospital's use could be grown, and this became the first recorded garden on the site.

By 1789, the Monserrate estate was being rented by Gerard de Visme (1725–c.1795), a wealthy English merchant from a Huguenot family, who built a neo-Gothic house with a large garden on the site of the by-then ruined chapel. From 1794 onwards, he sublet the property to the British aristocrat and writer, William Beckford (1760–1844), who had recently been involved in a sex scandal and was seeking refuge away from England. Beckford – who had inherited vast wealth,

BELOW: The *Palácio de Monserrate*, designed for Sir Francis Cook in 1856 by the English architect James T Knowles, combines Moorish, Arabic and Romantic styles.

and his English country seat, Fonthill Abbey in Wiltshire – was the main reason for Monserrate's early fame, although he didn't actually design or build anything himself while there.

Cook's Project

By 1856 not only the house but the gardens were in ruins when they were bought by an English textile millionaire, Sir Francis Cook (1817–1901). Cook saw Sintra as the perfect place for a summer home, and, using the ruined earlier building as a starting point, the architect James Thomas Knowles (1806–1884) built it back up with a blend of Arabic, gothic and Indian architectural styles.

Cook brought in the landscape designer William Stockdale, the botanist William Neville and head gardener James Burt and began to fill the Garden with exotic plants from across the world. They laid the foundation of the Garden we see today. Cook's ambition was to make a garden at Monserrate that would rival the greatest botanical collections of Europe. He knew William and Joseph Hooker, father-and-son directors at Kew, and established an exchange of plants with them. Dr Gerald Luckhurst, the landscape architect and garden historian who has led the restoration of Monserrate since the 1980s, says that Monserrate contained a greater variety of plants than any other garden in Europe, except for Kew. The Garden still retains the first lawn planted in Portugal, with an unusual cambered surface; this is also the oldest irrigated lawn on the Iberian Peninsula.

Early articles written about Monserrate provide important insights into the provenance of its plants. In 1885, C A M Carmichael wrote in *The Gardeners' Chronicle*:

'... palms and cycads are mingled together with the singular vegetation of Mexico. Tree Ferns elbow Camellias, Ipomeas and Bougainvilleas cover entire walls and Tacsonias overpower tall trees in their rank luxuriance.'

As Carmichael's list reveals, Monserrate's plants were coming from all over the world, including North America, Mexico, Brazil, Chile, Japan, China and South Africa. A guide to the Garden dating from 1923 and written by then head gardener Walter Oates (who came here from La Mortola) shows that the Garden was very much on the tourist route, and also usefully lists plants from that time.

RIGHT: An aerial image of the renovated Palace of Monserrate, showing its wonderful high setting in Sintra.

Literary Inspiration

Lord Byron visited Monserrate in 1809, but by then the house had been abandoned after being sacked by Napoleon's troops. The estate's romantic appearance was a source of inspiration for Byron, who used the setting in his epic poem 'Childe Harold's Pilgrimage', which was serialized from 1812. This put Monserrate firmly on the tourist route, and as a result, it enjoyed a new spell of fame.

The Garden today

The Garden has undergone a long restoration, starting in the mid-1980s with a team headed by Dr Luckhurst, a landscape architect and garden historian. The Palace, which was emptied of furniture and collections in the middle of the 20th century and virtually abandoned for fifty years, has also been renovated.

Visitors should bring good walking shoes as the Garden has a rocky terrain with steep walks. Exploring it is an experience to remember – there are waterfalls, many indigenous plants from the region (such as strawberry trees, holly bushes and cork trees), araucarias and palm trees, tree ferns from Australia and New Zealand, and cinnamon trees from South India; as well as a Japanese garden with Asian plants, bamboos and camellias, and misty humid valleys full of azaleas and rhododendrons.

The Valleys

The three main valleys, divided into zones relating to their microclimates, are planted with specimens from regions and countries representing the best horticultural match. In one valley is a Rose Garden, which has undergone an incredible restoration under Dr Luckhurst. Original paths, flower beds and even some of the original roses have been rediscovered and revitalized. Different varieties of rose bushes all predating 1920 have been planted, with staggered periods of flowering, including some roses documented as having previously existed in Monserrate. They also have more recent rose varieties, of Indian origin. These are hybrids of *Rosa gigantea* – the parent of the Portuguese roses, such as 'Bella Portuguesa', raised in the early twentieth century at Lisbon Botanic Garden. In Fern Valley is a wonderful collection of tree ferns, while the Mexican Garden contains plants from hot climates such as agaves and yuccas, in this, the hottest and driest area of the Garden.

The biggest tree in Monserrate is a Norfolk Island pine measuring over 50 metres (164ft). The palm collection includes some 70 different taxa, including 24 that survive from the original plantings. A large lake holds a collection of exotic aquatic plants, although no longer any *Victoria amazonica*, which proved particularly challenging to maintain.

OPPOSITE: Built as a ruin, the remains of the church are now engulfed by an Australian rubber tree.

ABOVE: The Mexican Garden at Monserrate has recently been restored. Plants include palms, yuccas, agaves and cycads.

Jardim Botânico do Rio de Janeiro
Rio de Janeiro, Brazil (1808)

Rio's Botanical Garden, dating from 1808 and the first such garden in Brazil, inhabits a 54-hectare (133-acre) site in the south of the city, at the foot of the Corcovado Mountain in the shadow of the famous 'Christ the Redeemer' statue. The Garden stands as a monument to Brazil's rich diversity of flora and fauna, containing over 6,000 species of tropical and subtropical plants and trees, including some 900 varieties of palm trees. 40% of the site is cultivated: the rest is Atlantic Forest (*Mata Atlântica*) spreading up the mountain slopes. It is home to colonies of Capuchin monkeys and marmosets, along with over 140 species of birds, including toucans, rusty-margined guan, slaty-breasted wood rail and white-necked hawks.

The History

The Garden was established by John, the Prince Regent of Portugal and later King John VI of Portugal and Brazil. It was founded to function as a nursery where plants imported from the West Indies, such as pepper and cinnamon, could acclimatize. A location was chosen low down on the southern flank of the Serra do Corcovado at the edge of the Rodrigo de Freitas Lagoon. Shortly before, the Portuguese authorities had built a large gunpowder factory here (which today is one of the Garden's several museums). João Gomes da Silveira Martins, the future Marquis of Sabará, was placed in charge of the Garden, which was at this time more of a Royal Kitchen Garden, its collections boosted by plants seized by Portuguese customs officials and, after

1814, by the introduction of Chinese tea plants. In 1819, the Regent Dom Pedro I rebranded it the 'Royal Botanic Garden' and it was opened to the public three years later.

Charles Darwin

One notable visitor in May 1832 was Charles Darwin, the father of evolutionary theory, then in the early stages of his legendary voyage aboard the *Beagle*. He saw specimens of camphor, sago, cinnamon, cloves, breadfruit, mango and jackfruit here and wrote in his *Voyage of the Beagle*: 'One day I went to the Botanic Garden, where many plants, well known for their great utility, might be seen growing.' However, a more candid entry in his personal diary suggests he was underwhelmed by some of what he saw, not least the Garden's collection of tea trees. 'I felt quite disappointed at seeing an insignificant little bush with white flowers & planted in straight rows,' he wrote. 'Some leaves being put into boiling water, the infusion scarcely possessed the proper tea flavour.'

The Avenue of Palms

One of the Garden's iconic sights is the Avenue of Royal Palms, which stretches for 750 metres (about half a mile) from the entrance into the centre of the site. It consists of 134 Cuban imperial palms 30 metres (98 feet) tall, each descended from a single antecedent, the Palma Mater, which was sadly destroyed by lightning. The Palma Mater itself originated as a cutting in a box of plants intercepted by officials on its way from

Mauritius in the early years of the Garden. It was said to have been planted by Prince Regent John himself, and created the fashion for imperial palms that spread across Brazil in the 19th century. For those intimidated by the sheer scale of the Garden, it is possible to pick up an electric golf cart at the top end of the avenue.

ABOVE: The 30-metre (98-foot) statue of Christ the Redeemer on top of the Corcovado mountain overlooks the *Jardim Botânico do Rio de Janeiro*.

Pau brasil

The Garden boasts a collection of *Pau brasil*, the species from which the country gets its name and which is today recognized as its national tree. From the early 16th century, it was harvested by the Portuguese for the distinctive red oil produced in its trunk that could be used for dyeing fabrics. Its durable and unusually flexible wood is also highly prized, not least by musical instrument manufacturers who use it for making violin bows.

ABOVE: An aerial view of the Garden shows the Friar Leandro do Sacramento Memorial, dedicated to the first director of the Garden.

OPPOSITE: The 750-metre (2,460-foot) Avenue of the Palms. The stately giants are all descended from the same Palma Mater.

LEFT: The red bridge of the Japanese Garden. Plants from Japan include a collection of cherry trees and skilfully trained Bonsai trees.

OPPOSITE: One of the six lakes with various species of papyrus and lotus including the giant *Victoria amazonica* (here known as *Água-Pé* or *Vitória Régia*).

Japanese Gardens

The Japanese Gardens are resplendent when their cherry trees are in blossom, while a collection of expertly cultivated Bonsai trees are exceptional. The Gardens are thoughtfully laid out, with a series of bridges providing ideal spots to glimpse the carp and Japanese Koi that pack the ponds. The *pièces de resistance*, though, are the six lakes brimming with various species of papyrus and lotus including the giant *Victoria amazonica* (here known as *Água-Pé* or *Vitória Régia*).

The Garden today

With Brazil confronting the challenges of rainforest deforestation, the Garden prioritizes work towards conservation and ecological restoration. It is recognised as a UNESCO biosphere reserve, where locals are encouraged to work with biodiversity and to explore the sustainable use of environmental resources. The Garden's densely packed Amazon section is home to rainforest species including rubber, cocoa, and capirona (*Calycophyllum spruceanum*), a tree that changes colour with the seasons. There is a Herbarium with approximately 330,000 reference specimens, along with Brazil's largest botanical library containing more than 32,000 volumes.

Meanwhile, the Orchidarium, based in an iron and glass house remodelled in the 1930s, contains more than 3,000 specimens of over 500 different orchid species, and functions as a research centre spearheading new discoveries into these diverse plants.

Among the Garden's unexpected delights are a lake full of *Victoria amazonica* water lilies, usually only seen in the Northern Hemisphere in glasshouses but here flourishing outdoors in their natural habitat. There is also a Sensory Garden, designed with visually impaired people in mind with aromatic plants and Braille signage.

Royal Botanic Garden Sydney
Australia (1816)

The setting of Sydney's Royal Botanic Garden is perhaps the most stunning of any in the world, overlooking Sydney Harbour and next door to the Sydney Opera House. The oldest botanic garden and scientific institution in Australia, its 30 hectares (74 acres) contain 45,000 plants from 10,000 plant groups, arranged over 17 themed gardens. It attracts some three million visitors annually, and was granted its 'Royal' title in 1959 after a visit by Queen Elizabeth II.

The History

The Garden sits on the site of the traditional lands of the aboriginal Gadigal people, who knew it as Woccanmagully and considered it a place of sacred and ceremonial importance, as well as a fruitful hunting ground.

European settlers established a farm here (at what they called 'Farm Cove') in 1788. It failed, but its land was to be used as a later estate, founded by Governor Lachlan Macquarie in 1816 as part of the Governor's Domain. Macquarie was to transform the land into an English Gardenesque-style landscape: he enclosed the Domain with a sandstone wall, making rules that no cattle or animals should be allowed on the land.

In the same year, Scotsman Charles Fraser – a soldier, explorer and gardener – arrived aboard a convict ship, the *Guildford*, and was appointed Superintendent of the Garden. He collected plants on field trips and expeditions, developing the Garden along scientific lines, and Macquarie gave him the title of Colonial Botanist of New South Wales. Fraser was to begin the sharing of specimens, plants and seeds, in particular with the *Jardin des Plantes* in Paris but also with other botanic gardens, establishing a scientific network that continues today. He created an experimental garden where he made trial plantings of newly-discovered Australian rainforest trees, and the first grapevines introduced to the colony. In 1831, Fraser opened the expanded Garden to the public.

The Cunninghams

Trained at London's Kew Gardens, King's Botanist Allan Cunningham arrived in Sydney in 1816, and he spent fourteen years plant-hunting and exploring in Australia, sending his finds back to Kew. Cunningham was given the title of 'King's Collector for the Royal Garden at Kew', and he was so successful that a hothouse at Kew was renamed 'Botany Bay House' to hold his specimens.

After the death of Fraser in 1831, Cunningham was offered the Superintendent's position in Sydney, but instead put forward his brother, Richard, a fellow botanist and Kew alumnus. However, when Richard died on a plant-hunting expedition in 1835, Allan Cunningham finally accepted the position. Gardens like Sydney were thought of as staging posts for new plants that could prove valuable to the

RIGHT: An aerial view of Bennelong Point, the peninsula where Sydney's Opera House and Royal Botanic Garden are located.

British Empire. These would be taken back to England for testing at Kew, and then sent out to the colonies. The brown beech *Pennantia cunninghamii* and the rainforest featherwood *Polyosma cunninghamii* were both named in commemoration of the brothers.

Charles Moore

One of the Cunninghams' successors, Charles Moore, arrived from Kew in 1847, and he was appointed government botanist and director of the Gardens (a post he held until 1896). Like his predecessors, he undertook plant and seed-hunting expeditions, and was particularly interested in the native flora, so different to what he knew. He founded many important research areas at the Garden – a Herbarium, a library and a lecture theatre – and added a new area of medicinal plants. Moore's brother

was director of the Botanic Gardens at Glasnevin in Dublin, and the pair set up botanical exchanges between the gardens.

The International Exhibition

The Southern Hemisphere's first-ever World's Fair, the Sydney International Exhibition, was hosted at the Botanic Garden in 1879. The Garden Palace, inspired by London's Crystal Palace, with a 64m/210ft high central dome that could be seen as far away as the North Shore, was constructed near where the Palace Rose Garden and Pavilion are today.

In 2018, Sydney opened a Southern Africa Garden – including 60 plant and 30 bulb species from Botswana, Lesotho, Namibia, South Africa, Swaziland and Zimbabwe – which draws on the African cycad plantings made by Moore in the 1850s.

OPPOSITE: Native and exotic plants are carefully cultivated to give a display of colour all year round.

BELOW: Palm Grove – the first palms were planted here in 1862, and there are now more than 300 species.

The Birthplace of Australian Wine?

It is thought that Australia's oldest wine grapevines could have originated in the Garden. In the 1830s, vine cuttings arrived on a convict ship, the *Camden*; these came from France and Spain, and had been collected by James Busby. They were first established in the Botanic Garden before being sent out to the Hunter Valley, which later became a centre of Australian viticulture.

Palm Grove

Palms survive here that were planted by Moore, as well as some of the early rainforest trees collected by Fraser, including the Weeping Lilly Pilly (*Waterhousea floribunda*), Hoop Pine (*Araucaria cunninghamii*) and Yellowwood (*Flindersia xanthoxyla*). There are now three hundred species of palms, including many rare ones, and there are small groves of Australian fan palms (*Licuala ramsayi*). The Garden's tallest tree is a Queensland Kauri Pine (*Agathis robusta*, 1853) which was collected by John Bidwill in 1849.

ABOVE: The iconic 17-metre (55-foot) high steel and glass Pyramid opened in 1972. It was demolished in 2015 and replaced by the Calyx.

RIGHT: The Calyx is described as 'the sepals of a flower, typically forming a whorl that encloses the petals and forms a protective layer around a flower bud.' It is used as an exhibition space within the Garden.

The Herbarium

The National Herbarium of New South Wales holds over a million preserved plants, including some collected in 1770 by the botanists on James Cook's HMS *Endeavour*, Joseph Banks and Daniel Solander (although the Herbarium at Kew holds many of the earliest collections sent from Australia during the colonization period). Vital for the study of Australian native plants and to inform conservation work, the collection has so far digitized 1.4 million botanical specimens (including over 800 from Banks and Solander). The largest project of its kind in the Southern Hemisphere, it allows researchers worldwide to explore high-resolution digital images online, protecting the specimens from handling.

The Pyramid House

Designed by E H Farmer and completed in 1972, this 18-m/59ft-high pyramid-shaped glasshouse was the first of its kind to have an automatic heating and cooling system to control humidity. To celebrate the Garden's 200th birthday in 2016, it was replaced by a new glasshouse on the same site, the Calyx, which houses the largest interior green wall in the Southern Hemisphere.

Tresco Abbey Garden
Isles of Scilly, England (1834)

A haven for 20,000 plants from over 80 countries, this 19th-century garden is found on Tresco, one of the Scilly Isles, 45 kilometres (28 miles) off the Cornish coast of England. A subtropical island, blessed with white sand and turquoise sea, it is home to just 150 people. Tresco has its own climate due to the Gulf Stream, making it possible to grow rare and exotic species not seen elsewhere in the open in the UK. Described as 'a perennial Kew without the glass', the Garden is designated Grade I in the *Register of Historic Parks and Gardens of Special Historic Interest in England.*

The History

In the 1830s, merchant banker Augustus Smith bought a long lease on the Isles from the Duchy of Cornwall and chose Tresco, the second-largest island in the archipelago, as his home. He built a house next to the ruins of a 10th-century Benedictine abbey and a 12th-century priory, near a fresh water pool and overlooking the sand dunes and beach at Carn Near. Within the abbey ruins, he cultivated a garden, creating a series of terraces on its steep slope.

To defend against the wild Atlantic winds, he built a series of high granite walls, planted gorse and erected a shelter belt of imported trees from North America, including Californian Monterey pines and Monterey cypresses. Today, an incredible hedge of oak (*Quercus ilex*) 15.25 metres (50 feet) high protects the plants.

Smith's descendant, Arthur Dorrien-Smith, oversaw a large expansion of the Garden in the early years of the 20th century. A prolific traveller, Dorrien-Smith undertook expeditions to, for example, South Africa, New Zealand and Australia, returning with over 2,000 new specimens.

The devastation of natural forces

Tresco suffered terribly during the late 1980s, when an exceptionally long cold snap – reaching as low as -25°C – killed 80% of the Garden. The notorious 'great storm' of 1987 and further gales in 1990 decimated the trees of the shelter belt. Faced with such devastation, the (now) curator Mike Nelhams and (now) head gardener Andrew Lawson called upon their network of contacts to repopulate the Garden virtually from scratch. Their botanic connections stretch to Italy, and the gardeners have had many visits to the University of Genoa's *Giardini Botanici Hanbury* (Hanbury Botanical Gardens) at La Mortola.

Climate

Its proximity to the sea and the elements could lead to horticultural disaster, but the Garden has been thoughtfully planted to take advantage of the natural terrain, the soil and the weather. Both maritime and Mediterranean plants thrive in poor soil and so are very happy here. The

RIGHT: The gardens at Tresco were made around the ruins of a Benedictine priory dating from the Middle Ages. Parts of the granite walls of its nave and chancel and two arches survive.

Garden has 6.8 hectares (17 acres) of dry-temperate-to-wet-temperate areas, and often the long dry spells in summer can last ten weeks, with no rain and no irrigation system in the Garden. The gardeners tend to do their planting in the winter as they don't get a frost.

The seasons also start earlier here with spring flowers weeks ahead of the mainland. In the 19th century this led to a competition between Tresco Abbey Garden and the glasshouses of Kew. Each New Year's Day they would count the individual species in flower. This still takes place, and usually around 300 species will be found to be flowering.

Head gardener Andrew Lawson comments: 'The flower count is a wonderful island tradition that has been taking place since the Garden's early years when Joseph Hooker of Kew Gardens and Augustus Smith of Tresco competed for plant species on New Year's Day.'

Winter flowering

There are over 100 varieties of camellias at Tresco, flowering from November. The temperate range includes subtropical Aloes from South Africa that bloom

ABOVE: 19th-century ship figureheads from the Valhalla Collection.

OPPOSITE: Plants at Tresco survive long summers with no rain or irrigation.

The Valhalla Collection
Smith also gathered an impressive array of 19th-century figureheads salvaged from the islands' shipwrecks. Homed in an open-fronted building, they are decorated with two tonnes of sea shells given by German sailors, and displayed as they were in Smith's time.

around Christmas time. On the top terrace are Leucadendrons and Protea, members of the Proteaceae plant family endemic to South Africa, and Banksias from Australia, which flower from December onwards. The *Banksia marginata* (Silver banksia), to be found in Australian heath areas, was used by 19th-century Aboriginal people to strain drinking water, while its woody stems were made into tools and also woven into baskets and mats.

Signature plants

Tresco exhibits some of the biggest and most fascinating bromeliads, including the *P. alpestris* (with its blue flower spikes) and the even showier and taller *P. berteroniana*. In the Lower Garden, which enjoys shade and protection from the wind, is a large population of plants from the New Zealand rainforest, including numerous ferns.

Another tall plant is the Giant Dandelion, originally from the Canary Islands, a self-seeding plant, related to the common garden dandelion but giant-sized – it can grow to more 3.65 metres (nearly 12 feet)!

The Flame trees are incredibly happy on Tresco, while their intolerance to frost and cold winters mean that they are not grown elsewhere in the British Isles. They flower from May to July with clusters of bright red flowers – the vivid scarlet hues are so bright they be can seen from the other islands. According to Mike Nelhams: 'This is probably the best tree in the world for wind and salt spray

ABOVE: Tresco is a subtropical island, about 45km (28 miles) off the Cornish coast, with a climate where exotic plants flourish.

RIGHT: Aeoniums in the Mediterranean Garden frame the statue of Gaia, Goddess of Earth, sculpted by David Wynne in 1989.

tolerance, making it quite at home on the islands.'

In the eastern part of the Garden, you'll find the Silver tree (*Leucadendron argenteum*), native to the South African Cape and covered in fine, tiny hairs that resist salt and wind. The Norfolk Island pine is another tree that does well here, producing a new layer of branches each year.

Arguably the Garden's best-named specimen is the Robinson Crusoe Cabbage Tree (*Dendroseris litoralis*). Decorated with beautiful orange flowers, it comes from Robinson Crusoe Island, one of the Juan Fernández Islands, 643km (400 miles) off the coast of Chile.

Royal Botanic Gardens, Kew
London, England (1840)

The Royal Botanic Gardens in Kew are thought to be the largest botanical gardens in the world, a 121-hectare (300-acre) site on the edge of London, home to the world's biggest collection of living plants and more than 8.5 million preserved plant and fungal specimens. It is a setting rich in history that spans from royal follies to colonial exploitation and wartime bombing. Kew states its mission is to 'protect plants and fungi for the future of all life on Earth', and to investigate the plant world for new sources of food, medicine, fuel and materials.

Kew continues to help us understand the wonders of nature and to question how we care for it in a world where climate is now dangerously being ignored. Installations like The Hive whirring away in the middle of a wildflower meadow, recreating life inside a beehive, remind us of the challenges bees face to survive and the landscape they and we need to thrive. Designed by artist Wolfgang Buttress, it features a thousand LEDs lighting up according to the vibrations of bees, and creates a musical symphony in the key of C – the same key as bees buzz in!

The History
Grade I-listed Kew Palace, the smallest of all the British royal palaces, was originally built for a silk merchant in 1631. In the 1720s George II and Queen Caroline arrived with their three eldest daughters, to use Kew and nearby Richmond Lodge as weekend retreats, enjoying some privacy away from Court life.

It was the arrival, in 1736, of Princess Augusta of Saxe-Gotha, who came to England aged 17 for her arranged marriage to Frederick, Prince of Wales, that was to change the landscape. Frederick took a lease on the Kew estate, and both he and Augusta became involved with reshaping it. Frederick began a collection of exotic plants, and after his sudden death in 1751, the now Dowager Princess of Wales Augusta created a 3.6-hectare (9-acre) botanical garden just south of the Orangery. By 1768, it contained 2,700 species, and botanist and gardener Thomas Knowlton said that the Princess's gardens had one of the best collections in the kingdom, if not the world.

As a reminder of Augusta's garden, an area of Kew is reserved for a collection planted out in strict Linnaean order and with each plant labelled. The Princess of Wales Conservatory is also named for her.

18th-Century Follies
Augusta commissioned various buildings and follies from architect Sir William Chambers, including the Pagoda, the Orangery and the Ruined Arch (all survive), and the Great Stove, the Temple of the Sun and the Mosque (not extant). A recent restoration of the Ruined Arch revealed that it contains pieces of Greek and Roman sculpture.

George III, who reigned from 1760, inherited both the Richmond and Kew

RIGHT: An autumnal aerial view of the Royal Botanic Gardens at Kew, London, showing the Palm House and lake.

ABOVE: A spring display of bedding in the William Nesfield parterre of 1848, in front of the Palm House.

OPPOSITE: The Garden of Peace, based on a traditional Japanese Tea Garden, at Kew.

A Series of Calamities

Undoubtedly the Gardens' unluckiest (or perhaps luckiest?) tree is a *Pinus nigra*, which has twice been hit by lightning as well as by a small, low-flying aircraft that crashed in 1928 (a second aircraft crash-landed in the Gardens ten years later). Other catastrophes included the burning down of the tea pavilion by Suffragettes in 1913, and thirty hits by Second World War bombs. The Gardens once had their own fire brigade and still have their own police force, one of the smallest in the world.

estates, joining them together and creating Kew Gardens (plural for the two estates). On the advice of Sir Joseph Banks, to support voyages of discovery to new lands, the King commissioned a joint Royal Navy/Royal Society scientific expedition to the South Pacific Ocean. Lieutenant James Cook was to command HMS *Endeavour*, and Banks was to join the team of botanical explorers on board. During the voyage (1768–1771), the botanists collected over 1,000 species of plants new to science, and the seeds were sent to Kew.

On Banks' return to England in July 1771, he became Kew's first unofficial director. Under his guidance, it became a central hub collecting seeds and specimens, and he built an international exchange of resources making it one of the pre-eminent botanical gardens in the world. Banks was also behind the practice of sending out plant hunters/collectors, including Francis Masson, a Scottish gardener and Kew Gardens' first official plant hunter – who, in 1772, visited what is now called the Floral Region in South Africa, and returned with thousands of plants.

By the 1840s the Gardens were in a state of decline, and after a damning report it was decided that they be handed over to the State. Kew had been left behind by the Colonial Gardens which were reaping benefits across the globe. Set up in Kew's image they were forging ahead with their focus on economic botany and science.

A new director, Sir William Jackson Hooker, began its revival, establishing museums, the Herbarium and Library, and a Department of Economic Botany.

The Palm House
Hooker commissioned the Palm House, designed by Decimus Burton and Richard Turner, and built between 1844–48. A living scientific laboratory, the Palm House holds many rare tropical and subtropical plants, and some that are now extinct in the wild.

In this glasshouse is the oldest pot plant in the world, the Eastern Cape giant cycad (*Encephalartos altensteinii*): it weighs more than a tonne and measures over 4 metres (13 feet) in height. Brought to Kew in 1775 by Francis Masson from South Africa, the plant has only ever produced one single cone during its time here. This was in 1819, and the event was witnessed by Sir Joseph Banks.

A miniature railway runs underneath the Gardens' lawns to the Palm House, designed to transport coke and ashes to and from the boilers, with flues used to channel the smoke to a chimney disguised by Burton as an Italian campanile.

The Temperate House

This house, twice as big as the Palm House, was built from 1860–99 for Kew's largest plants. Now filled with 10,000 individual temperate plants, its 1,500 species, from Africa, Australia, New Zealand, the Americas, Asia and the Pacific Islands, all require conditions above 10°C to survive.

Under Hooker, and his son Sir Joseph Dalton Hooker who also became director, Kew corresponded with over thirty other botanic gardens growing plants of economic value. In the 19th century much of the work at Kew focused on the movement of valuable plants around the British Empire for agriculture and trade: this involved the confiscation of land from indigenous peoples, and the use of slavery. Sir

William Thiselton-Dyer, director of Kew from 1885 to 1905, wrote that 'we at Kew feel the weight of empire more than they do in Downing Street'. Kew now seeks to address its controversial past, pledging to 'decolonize' its collections by acknowledging their 'exploitative and racist legacies.'

The Aroid House

Designed by John Nash in 1825–30 to house tropical rainforest specimens, the Aroid House was originally one of two pavilions from the Garden Façade at Buckingham Palace. It was moved to Kew in 1836 on the orders of the King and houses the Titan arum, *Amorphophallus titanum*. A Titan arum flowered for the first time outside its native Sumatra at Kew in 1889. When a Titan arum bloomed for the second time at Kew in 1926, the crowds it drew were so big that the police had to be called to control them. The Aroid House has other specimens from the genus, some which are both rare and vulnerable to extinction.

OPPOSITE TOP: One of the octagonal wings of the Decimus Burton-designed Temperate House at Kew Gardens.

LEFT: The Waterlily House at Kew was originally built especially for the giant *Victoria amazonica* lily, but now it is the home of a smaller lily, *Victoria cruziana* and other tropical aquatic plants.

Royal Botanic Gardens Victoria
Melbourne, Australia (1846)

The Royal Botanic Gardens Victoria (RBGV) is a beautiful landscaped garden that has been part of the cultural centre of Melbourne for almost 180 years. The Gardens have been used for recreation and enjoyment just as much as scientific research and conservation. There are over 38 hectares (94 acres) of land, featuring no less than 12,000 different species of flora from around the world (more than 51,000 individual plants), placed in 31 collections. The Royal Botanic Gardens Melbourne name was changed in 2015 to Royal Botanic Gardens Victoria, incorporating Melbourne Gardens, Cranbourne Gardens, the National Herbarium of Victoria and the Australian Research Centre for Urban Ecology (ARCUE).

Melbourne Gardens History

These gardens were established in 1846 beside the Yarra river in the heart of Melbourne, on land belonging to the Kulin Nation, by Lieutenant-Governor Charles La Trobe. The RBGV is a former outpost of Kew Gardens in London, and amongst other histories has a colonial past and also one of displacement of indigenous peoples: this is a difficult subject that the Gardens are now addressing.

In 1857, the German-Australian botanist and physician Ferdinand von Mueller (1825–1896) was appointed as their Director. He was already the government botanist for Victoria, a post that had been created for him. Mueller was to become one of the most acclaimed botanists of the 19th century, and was ultimately awarded knighthoods by over twenty countries, including a Papal one. Mueller established the Gardens' scientific centre, the National Herbarium of Victoria, and did not regard the creation of a beautiful landscape as an object in itself. It seems that maybe this was the reason for his dismissal from his post, after complaints about the aesthetics of the Gardens had been made by numerous important Melbourne figures. Mueller saw this as an injustice, and of all his achievements at the site he was said to have been proudest of the geyser in the lagoon, the flowering

---

The Aboriginal Heritage Walk

Today, the Royal Botanic Gardens gives Aboriginal Heritage walks of the important cultural sites of the local Kulin Nation, with First Peoples guides, where one can learn about Aboriginal plant uses and customs. The RBGV aims to highlight the histories and living cultures of the traditional owners of the land, an example being the Oak Lawn, which was established throughout the 19th century with species of oak from Europe, Asia, and America, planted by early colonizers on the land of the Wurundjeri people.

RIGHT: The iconic Cockscomb Coral Tree (*Erythrina crista-galli*) originating from South America, near the Observatory Gate of the Royal Botanic Gardens Victoria.

of the South American giant waterlily, *Victoria amazonica*, and the Living Plant Collection. He was replaced by William Guilfoyle (1840–1912), an English landscape gardener and botanist.

Guilfoyle was to rearrange Mueller's plants and create an '18th-century English landscape' style, with sweeping lawns and follies such as the classical 'Temple of the Winds': it was to be an idealized vista, with views from a series of fixed points. He established the Fern Valley, the Australian Border – and, after 1897, when the course of the Yarra was straightened, he transformed the lagoon into a landscaped lake.

The Botanic Gardens as a place of study and recreation

During the 19th century, the scientific and economic interest of plants became entwined with the use of exotics in the gardens of Britain and European countries and in their colonies. Botanic gardens were also seen to have a role in educating the public. The 1871 Board of Enquiry into the Melbourne Botanic Gardens stated: 'Such a Garden would have more than a scientific object – it should also be a place where the whole colony could study horticulture, arboriculture, floriculture and landscape gardening in their most perfect forms – it should especially be a model of careful and thorough cultivation, of well-planned scientific effect, and of art skilfully applied to the embellishment of nature.'

One example of this 'embellishment of nature' was the introduction of foreign species to the Australian gardens. Some of these were mistakes of the colonial scientific age – plants such as the European Blackberry (*Rubus fruticosus*). It was first introduced into Australia

OPPOSITE: The bustling city of Melbourne, Australia's second-largest, can just be glimpsed beyond the quiet and calm of the Royal Botanic Gardens.

BELOW: On the highest point of the Gardens' Guilfoyle's Volcano, the now ornamental reservoir planted with cacti and succulents, originally made in 1876.

in the 1830s, and Mueller, together with the first Curator of the Gardens at Melbourne University, Alexander Elliot, recommended that this blackberry be planted to control soil erosion along creek banks. It became a fashion of early settlers to use blackberry for erosion control, and combined with the plant's invasive potential, this soon produced a serious weed problem. Today in Australia the blackberry is considered a significant threat to biodiversity, and it affects almost 9 million hectares of grazing land.

The Living Collections

The Royal Botanic Gardens have specially curated Living Collections. These are made up of diverse plants from across the world, including rainforest flora, cacti and roses. There are also Californian species, plants from Southern China, and, in the Rare and Threatened Species Collection, plants from southeastern Australia.

The National Herbarium of Victoria

The Herbarium houses the internationally important State Botanical Collection, crucial for scientific research in botany, taxonomy, mycology and conservation. It has 1.5 million preserved plants, algae and fungi, and is the largest herbarium collection in Australia, with research being focused on native Victorian and Australian flora.

The Library of the Herbarium is one of Australia's most comprehensive botanical book collections, and includes the large, private library of former Director Ferdinand von Mueller.

Cranbourne Gardens

As the Botanic Gardens became focused on the problems caused by climate change, scientific research, education and conservation concentrated on native areas and indigenous plants. With this in mind, a satellite or sister garden was established in 1970, when land on the edge of Melbourne's city limits, originally

ABOVE: The National Herbarium of Victoria was founded in 1853 to house the State Botanical Collection.

OPPOSITE: The Gardens are a living laboratory. The size and health of trees has been monitored here, using high-resolution GPS, by University of Melbourne researchers.

inhabited by the Boon Wurrung people, was acquired by the Gardens for the purpose of establishing a contemporary botanic garden dedicated to endemic Australian bushland plants. The wild landscape of Cranbourne Gardens, with its native bushland, heathlands, wetlands and woodland, features over 100,000 plants from 1,900 plant varieties, and was opened to the public in 1989. One area, the Bloodwood Garden, recounts how Ferdinand von Mueller sent blue gum seeds to Rome as a means to dry the swamps outside Rome and eliminate malaria, for which he was awarded his papal knighthood.

Missouri Botanical Garden
St Louis, USA (1859)

The Missouri Botanical Garden is the oldest botanical garden in continuous use in North America. Founded in 1859, it is also known as Shaw's Garden after the founder and philanthropist Henry Shaw. It is recognized internationally for its scientific research, with over 4,000 live trees (some rare and some planted by Shaw himself). The Garden has been involved in the conservation of plants from native American regions and also from Madagascar, China and Central America. It includes almost fifty themed gardens with five formal gardens, and twenty-three demonstration gardens.

The History
Henry Shaw was born in Sheffield, England in 1800 and was educated at Mill Hill School in London. The school, located on the estate that once belonged to Quaker botanist Peter Collinson, contained rare species and interesting trees, including three Cedars of Lebanon said to have been planted by Carl Linnaeus, of which Shaw was almost certainly aware.

Having already visited North America on business with his father, in 1819 he returned to begin a new life in St Louis, Missouri. After making his fortune from hardware and real estate, he retired aged 39 and travelled extensively, mainly in Europe. In 1851, he visited the Royal Botanic Gardens and the Great Exhibition in London and saw the gardens of the Chatsworth estate in Derbyshire. He returned to St Louis with the intention of creating his own Kew Gardens. He corresponded with William Hooker, the

Director at Kew, who introduced Shaw to other important botanists, including the Harvard University botanist Asa Gray, and George Engelmann (who became his scientific adviser). They persuaded him to create both the Herbarium and the Library.

Henry Shaw's history as a slave-owner is acknowledged by the Botanical Garden, which states that 'it is our goal... to respect this history, educate, commemorate, and move beyond the past to improve the future.' Shaw died in 1889 and his mausoleum stands in a wooded grove on the property.

Next door to the Garden is Shaw's original 1850 residential estate, with an Italianate villa designed by the English architect George I Barnett. Shaw donated it to the city on the condition that it 'shall be used as a park forever': he believed that parks serve 'not only as ornaments to a great city, but as conducive to the health and happiness of its inhabitants and to the advancement of refinement and culture.'

The Garden
The Botanical Garden was designed with formal plantings, and is decorated with statues and pavilions mainly dating from the 19th century. James Gurney, Sr was Shaw's landscape designer, and he worked in the then fashionable British

RIGHT: The Climatron at Missouri seen from the main axis and the waterlily pools, showing the seven bronzes by Swedish sculptor Carl Milles which were installed here in 1988.

style of the Gardenesque. Previously, Gurney had been head of the aquatic plants section at Kew when the first *Victoria regia* (*amazonica*) had flowered. The flowering of the lily was such a phenomenon that it drew huge crowds, and even Queen Victoria visited.

The Herbarium, with more than 6.6 million specimens, is the second-largest in North America (the largest being at New York Botanical Garden).

The Climatron

The Climatron dome conservatory, built in 1960, was designed by T C Howard of Synergetics Inc. on principles established by the inventor R Buckminster Fuller. It was the first geodesic dome to be used for plants, and also the first air-conditioned greenhouse, with state-of-the-art climate controls. It has an innovative lighting system that allows

there to appear to be different time zones in the conservatory: it is noon daylight on one side of the dome and moonlight on other. The climate ranges from tropical rainforest, from the Amazon through to Hawaii and to India. The Climatron contains over 2,800 tropical plants, and in 1976 it was named one of the hundred most significant architectural achievements in United States history.

ABOVE: The Climatron is the first geodesic dome to be used as a conservatory for plants.

RIGHT: Inside the Climatron. It has no structural supports and has a state-of-the art air conditioning system which allows various climates to be recreated. It houses thousands of tropical plants from across the world.

Linnaean House

The Linnaean House, dating back to 1882, is the only greenhouse to survive from Shaw's time. It was designed by Barnett (who also created the Palm House and the Plant House in Tower Grove Park) as an orangery to hold the Garden's citrus trees and potted ferns and palms. The north section now contains a collection of camellias – including *Camellia japonica* 'Elegans', and two specimens of *Camellia japonica* 'Nobilissima' – originally planted in the 1930s. The Garden also features rare and endangered camellias, including *Camellia petelotii* var. *petelotii* (the Yellow Camellia).

Seiwa-en

This 'Garden of pure, clear harmony and peace' was designed in 1977 by the late Professor Koichi Kawana, a Japanese lecturer on environmental design and landscape architecture at the University of California. The 5.6 hectare (14-acre) Japanese Garden is one of the largest in North America, with authentic and symbolic features, minimalistic Japanese plantings, bridges, waterfalls and raked gravel. In April to May the cherry trees are in full bloom, with over forty specimens of Yoshino cherry, numerous weeping Higan cherry, Sargent's cherry and the Kanzan (*Prunus* 'Sekiyama').

ABOVE: The waterfall in the *Seiwa-en*, Japanese strolling garden, designed in 1977 by Professor Koichi Kawana.

Friendship Garden

The *Margaret Grigg Nanjing Friendship Garden* is modelled on 16th- and 17th-century 'scholar's gardens' in the southern provinces of China, and is designed to celebrate the longstanding scientific and cultural exchange between Missouri and various Chinese botanical institutions.

The Bakewell Ottoman Garden

In 2006, a new area of the Garden was laid out, inspired by the lost imperial gardens of the Ottoman Empire. The plantings consist primarily of native or naturalized Turkish flora, and of the citrus and fruit trees that were exotic introductions into those gardens; there are also fragrant roses and bulbous plants, including tulips. As St Louis and Istanbul lie roughly on the same 40° latitude, the Garden has a suitable climate for many of the same plants.

Conservation

The Garden is home to the Center for Plant Conservation (CPC), which works to save threatened species indigenous to the United States, and maintains the National Collection of Endangered Plants. Missouri scientists work with threatened plants both on site and in the wild, and the Garden is actively involved with thirty other gardens worldwide to restore natural habitats for threatened US varieties.

RIGHT: An autumnal view of the *Seiwa-en*. The garden was designed to be seen from various points around the 1.6 hectare lake; this view from the pebble beach is of one of the traditional Japanese bridges.

The Botanic Gardens of Singapore (various)
Singapore (1859)

Singapore has been described as the most biodiverse city in the world. It contains the Singapore Botanic Gardens – the first and only such place to feature on UNESCO's World Heritage List – as well as new and exciting botanic sites around the city. Two new gardens, the Jewel at Changi Airport and the Gardens by the Bay, are challenging the boundaries of garden design and technology.

The Singapore Botanic Gardens

The Gardens were founded at Tanglin in 1859, but the idea behind them came much earlier in 1822, when Sir Stamford Raffles, the British colonial agent and founder of modern Singapore, developed a 'Botanical and Experimental Garden.' Raffles had it planted with 125 trees, seeds of nutmeg and cloves that were so successful that they became the suppliers of the first spice plantations of Singapore. He was a keen naturalist and regularly went on trips to explore the tropical flora and fauna of the region.

A strange plant, *Rafflesia arnoldii*, with no leaves but the largest flowers in the world, is named after him. Similar to the Titan arum, it was found during an expedition to Sumatra organized by Raffles.

The Kew Influence

Today's Singapore Botanic Gardens were begun more as a pleasure park by the Singapore Agri-Horticultural Society, and were supported by the British colonial government before being handed over to them in 1874, when botanists were sent from Kew Gardens, along with a new superintendent, Henry James Murton. Under Murton, the garden expanded and progressed into a colonial acclimatization centre, where plants of economic value could be tested, eventually evolving into a serious research hub. Murton hired William Krohn, a zoologist, and until 1903 there were animals in enclosures throughout the Botanical Gardens. Interestingly, Raffles had kept a menagerie earlier: it featured (amongst other creatures) an elephant, two orangutans, a tiger and a sun bear. It is said that Raffles fed his sun bear champagne, raised him with his children in their nursery, and allowed the animal to sit at his desk.

'Mad' Ridley's Rubber Industry

In 1888 botanist Henry Ridley arrived from the British Museum to take charge not only of the Gardens but also of the forests of the Straits Settlements. Some years earlier in 1876, fifty rubber plants had been sent to the garden from Kew, but only five survived the trip and they died soon after. It was Ridley's arrival as Director that would kick-start the rubber-processing industry. He enthusiastically investigated the economic potential of the Pará rubber tree (*Hevea brasiliensis*) as a plantation crop. He had over 9,000 *Heveas* growing, and became so obsessed with his experiments that he was known as 'Mad' Ridley. He was so successful in his rubber cultivation that his techniques were used across Malaya, and the region became its biggest producer and exporter. Millions

ABOVE: An aerial view of the Gardens by the Bay nature park made in 101 hectares (249 acres) of reclaimed land in the Central Region of Singapore.

of hectares of Southeast Asian rainforest were lost as it was converted to the financially lucrative rubber groves, and this change in land use was to cause acute environmental problems in later years with the loss of biodiversity of the region.

The Orchid Boom

The Gardens started orchid breeding and hybridization programmes around 1928, drawing upon state-of-the-art *in vitro* techniques pioneered within their own laboratories. Today, the National Orchid Garden, to be found on the western side of the Gardens over three hilly hectares, contains more than 1,000 orchid species and 2,000 hybrids, a good proportion native to Singapore. Its VIP Orchid Garden includes examples of hybrids dedicated to a hundred or so celebrities, dignitaries and heads of state, honoured via the island's Orchid Diplomacy program.

Other attractions at the Gardens include a small area of tropical rainforest (which actually predates the Gardens) and the Jacob Ballas Children's Garden, named after its principal donor.

Singapore as a Garden City

Numerous new gardens in unlikely places within Singapore have been built in recent years. On the site of an old parking garage at Changi Airport is Jewel, a 10-storey-high glasshouse designed by Moshe Safdie's architecture firm: there are over 2,000 trees and the tallest indoor waterfall in the world. This is the Rain Vortex – at 40 metres high it pumps 500,000 litres of rainwater through the roof of the Jewel building. Safdie has said that architects need to design for climate change by making sure their structures can respond to more extreme weather and temperature conditions. The water is directed from the Rain Vortex to cool the climate-controlled indoor forest – created by Californian Peter Walker and Partners Landscape Architects with hedge maze and netted walkways inside the Jewel.

OPPOSITE: The Crane Fountain inside the National Orchid House, part of the Singapore Botanic Gardens UNESCO World Heritage site. There are over 60,000 orchid plants on display.

ABOVE: The Rain Vortex is the tallest indoor waterfall in the world, with a drop of 40 metres (131 feet). It can be found inside the Jewel Changi Airport, Singapore.

Gardens by the Bay

Another vision of the future is to be found at the Gardens by the Bay, made up of three waterfront gardens spanning 40 hectares (100 acres) next to the Marina Reservoir. This futuristic development was made on reclaimed, barren land. The idea for it came from Dr Kiat W Tan, botanist and former chief executive of Gardens by the Bay, who wanted to change the way we think of displaying plants. Using sustainable technologies, a team of international and local architects, engineers, and landscape specialists has transformed a desolate site into a much-needed verdant green space that is now home to a million plants from every continent except Antarctica.

Inside the Cooled Conservatories are the Flower Dome and the Cloud Forest, with a temperature at a moderate 23-25º Celsius on average. These lower temperatures allows a collection of plants that would not otherwise survive in the tropical climate of Singapore. There are eighteen giant tree-like structures: these are 'Supertrees', fifty metres tall, which serve as vertical gardens, with bromeliads, ferns, orchids and tropical climbers growing up and around them. Twelve of them can be found in the Supertree Grove: at night they are self-illuminated. Eleven of the 'Supertrees'

RIGHT: Evening light show of the Supertree Grove at Gardens by the Bay in Singapore. With cells on their canopies, the Supertrees collect their own solar energy for the evening light shows.

have built-in elements that enable them to be environmentally sustainable. Some have cells on their canopies which collect solar energy for the evening light shows; and others are fitted to act as air exhausts for the conservatories. They are powered by forage biomass, plant clippings from trees and shrubs on the island.

The Cloud Forest

Inside a greenhouse by Wilkinson Eyre Architects are the world's second tallest indoor waterfalls (after Jewel) at 35 metres (nearly 115 feet), and a lush mountain clad with plants from around the world. There are aerial walkways that allow close-up viewing of exotic plant species. The main conservatories, two glass biomes, are amongst the largest climate-controlled glasshouses in the

world. The aim of the Cloud Forest is to highlight the flora of those environments most likely to be affected by climate change. In the Flower Dome, the cool-dry Mediterranean zone; and in the Cloud Forest, they house a diverse collection of plants that are not commonly seen in this part of the world, some of which are of high conservation value.

Dragonfly & Kingfisher Lakes

The Gardens' lake system is an extension of the Marina Reservoir. It incorporates key ecological processes and functions as a living system, acting as a natural clean filtration system for water run off from the Gardens and providing habitats for plants and creatures such as fishes and dragonflies. Naturally treated water from the lake system is also used in the irrigation system for the Gardens.

OPPOSITE: The vertical gardens of the Supertree Grove stand 50 metres (164 feet) high.

ABOVE: The indoor waterfall at the Cloud Forest in the Gardens by the Bay is one of the tallest in the world. There are spiralling walkways which take you up and underneath the spray.

Bagh-e-Jinnah
Lahore, Pakistan (1860)

Lahore is known as a city of gardens with a long tradition of garden-making. In the period of British colonial rule, between 1849 and 1947, their style fundamentally changed, moving away from the historic *Mughal Charbagh* (a fourfold garden design relating to the four gardens of Paradise as related in the Quran) to become more focused on the cultivation of botanic collections. *Bagh-e-Jinnah*, set in the centre of the city, was an early manifestation of this development.

It is the only botanic garden in the world to incorporate a former Test cricket ground and English cricket pavilion.

History

The Garden was developed in 1860 as a Company Bagh of the British East India Company and was run by the Agri-Horticultural Society of the Punjab. It was known as Lawrence Gardens in honour of Sir John Lawrence, a veteran of the Indian Mutiny who was Chief Commissioner and then Lieutenant-Governor of the Punjab before serving as Viceroy of India from 1864 until 1869. The Garden was laid out by a gardener trained at Kew and modelled on the Royal Botanic Gardens there.

The first tree was planted in January 1862, as recorded by G R Elsmie, who had just arrived in Lahore to take up his duties as Assistant Commissioner:

One morning I accompanied Mr. Forsyth [then Commissioner of Lahore] to the newly laid out Lawrence [Garden], then little better than a barren wilderness... [and] planted the first tree of what was destined to become one of the most beautiful pleasure grounds in Upper India.

Soon, a large variety of trees were brought not only from different parts of India but from across the world, to be planted along the new Garden's avenues.

Its association with the scientific study of botany was consolidated when the Government College and Punjab University were established in 1864 and 1870, respectively. The Government College Botanical Garden was merged into the Lawrence Garden, making it one of the largest botanical collections anywhere on the subcontinent.

The Garden had a menagerie on its western side, and this land was eventually used to create Lahore Zoo in 1872 (one of the oldest in the world). By 1876 the Garden was continuing to be developed and had a collection of 80,000 trees and shrubs of 600 different species, from India, Austria, Syria and southern Europe.

The 20th Century

From 1912 onwards, the gardens were managed by the Government College, Lahore. However, a new botanical garden on a University campus site, inspired by the *Orto Botanico* at Padua, was to become the centre of botanical and horticultural interest. The Lawrence Garden became more of a nursery,

RIGHT: The *Bagh-e-Jinnah* in Lahore, Pakistan, previously known as the Lawrence Gardens, occupies about 57 hectares (141 acres).

growing plants and importing seeds and selling them to the general public.

Fruit trees were supplied from all over the British Empire; grapes and mulberries arrived from Kabul, peaches from Agra, and plantain from Calcutta (now Kolkata). An introduction by Mr A Hardie, the superintendent, to one of four catalogues published by the Garden in 1921 gives a sense of its operations: 'Arrangements have been made this season with Mr. Sutton & Sons of Reading, England to supply our entire stock of imported seeds, which has been selected for India.'

At the time of partition in 1947, the Garden took a new name, in recognition of Pakistan's founding father, Mohammad Ali Jinnah. It gradually built back its collections and today is home to 150 varieties of trees and 140 types of shrubs, along with creepers, palms, succulents and indoor collections. It has

ABOVE: The neo-classical Quaid-e-Azam Library, a public library inside the *Bagh-e-Jinnah* (Botanic Garden). The building dates from the mid-19th century, when Sir John Lawrence was Viceroy of India.

OPPOSITE: *Bagh-e-Jinnah* remains a popular place for local people to meet, with sports facilities, restaurants and an open-air theatre.

The Cricket Connection

There has been a cricket ground here since 1885, constructed as a distraction for the local British civil servants and government officials. It even hosted Pakistan's first (unofficial) Test match, a game against the touring West Indies side in 1948, becoming an official Test venue when Pakistan faced India in 1954-5, and the 3rd Test (in January-February 1955) was played at *Bagh-e-Jinnah*. Although now no longer a Test venue, it continues to host matches involving touring nations.

a particularly strong reputation for growing chrysanthemums.

The Garden Today

There are three nurseries that provide specimens to supply the Garden and support its research work, plus two libraries, the Dar-Us-Salam Library and the public Quaid-e-Azam Library dating from the 19th century.

The Garden is full of wildlife including unique Flying Foxes (Fruit Bats), which are being studied to understand how they are faring with climate change developments. The Global Climate Risk Index recently placed Pakistan fifth in its international rankings for vulnerability to climate change, and work is being conducted here on both fauna and flora to investigate the role that the Garden can take to improve knowledge and contribute to research into how plants and animals respond to alterations of weather and temperatures.

Jardín Botánico-Histórico La Concepción
Málaga, Spain (1865)

On the side of a steep hill, just a few kilometres north of the busy city and its harbour, is the subtropical and tropical Botanical Garden of Málaga. It has only been open to the public since 1994, but at its centre lie the remains of a family estate dating from the 19th century.

History

Originally, the site was the La Concepción hacienda, consisting of several farms along the Guadalmedina river, just north of Málaga. These grew vines, olive trees, almond trees and citrus orchards; records show there was a lemon grove here in 1750. Avocados, persimmon and pomegranate trees were added in the 19th century. La Concepción, built in about 1855, was the summer estate of Jorge Loring Oyarzábal and his wife Amalia Herédia Livermore: both were from wealthy families who had settled in Spain. They were to become the Marquis and Marquesa of the House of Loring, and had the idea of creating the Garden on their honeymoon, when they travelled around France, Germany and Switzerland, visiting many palaces and villas with wonderful gardens. They were especially interested in botanic collections, and on their return, they hired a French gardener, Jacinto Chamoussent, who selected and acclimatized exotic plants. He set up a system of hothouses in the Garden (these still exist), nurturing plants including pineapple (*Ananas comosus*) and coffee (*Coffea arabica*).

The Loring villa

At a high point in the Garden stands a large white classical villa, designed by German architect August Orth as a summer holiday home for the Lorings, and completed in 1865. It has wonderful Spanish hydraulic tiled floors, and a central patio with colonial-style wrought-iron work.

An arbour beside the villa also featured the new and fashionable wrought iron, which came from the family's own foundry. It was used for parties and dances, and local newspaper articles describe it as being decorated with Chinese lanterns. It still stands today, wound with climbing wisteria, and at Christmas when the Garden holds its 'The Botanical Lights' evenings the arbour is covered with thousands of fairy lights.

The museum and archeological remains

La Concepción had its own museum, the Museo Loringiano, built in the style of a Doric temple by another German architect, Wilhelm Strack. Opened in 1859, it housed the Lorings' growing collection of ancient artefacts. These included the *Lex Flavia Malacitana* bronze tablets dating from AD 81–84, on which are carved a set of Roman laws for the citizens of Malaca (as Málaga then was). Other pieces included a Roman mosaic, Roman statues, jewellery and coins, and even prehistoric artefacts. The best of these can now be seen in the National Archaeology Museum in Madrid and the Málaga Museum in the Palacio de la Aduana.

ABOVE: This 'historical viewpoint' with its *Mirador* gazebo, overlooks the city of Málaga, and was established here by Rafael Echevarría in about 1920.

The Echevarría Family, the Schoolhouse and the Mirador

In 1911, La Concepción was sold to Rafael Echevarría and Amalia Echevarrieta, a married couple from Bilbao. They expanded it and made changes to its design, adding streams, garden buildings and the Avenue of Palms. They also purchased contemporary sculptures and displayed them in the gardens.

The Echevarría family permitted more access to the estate, and built a schoolhouse there for local children and those of the estate workers: it still functions today as a classroom for visiting schoolchildren. They also added a feature that has become one of the Garden's defining elements: the *Mirador*. It is a belvedere set high on the hillside, with a domed roof and rectangular pool, placed to give the most wonderful views across the city.

La Concepción stayed in the Echevarría family until 1963. In 1990, after a period of abandonment and decline, it was bought by the Málaga City Council, and a restoration and preservation programme was instigated, with the objective of establishing a botanic garden with educational and scientific aims. The Garden has now been designated an Asset of Cultural Interest (*Bien de Interés Cultural*).

The plants

The Garden is a subtropical jungle with an outstanding palm collection, some plants dating from the time of the Lorings. One of the most impressive palms is a huge (3.5m/11ft tall) Chilean palm tree (*Jubaea chilensis*). The species – known as the Honey or Wine Palm, as it has sugary-sweet sap that makes an alcoholic drink when fermented – can live for up to 1,500 years. This one arrived in Spain from Chile in 1843. There is a large citrus orchard, laid out in terraces, and planted with native fruits such as Verna, a lemon with a good juice content grown widely in Spain. Lemons (*Citrus limon*) were exported from La Concepción to England in the early 20th century, each one individually wrapped in paper bearing the crest of the Garden.

There are various historic plants in the Garden, including an olive tree that is over 400 years old and measures

OPPOSITE: The *Museo Loringiano* or the Loring Museum, a Doric-style temple built in 1859 to house the Loring family's collection of Roman antiquities.

BELOW: The arbour, built 1863, was made from iron manufactured by Amalia Heredia Loring's fathers factory. Under the wisteria (*Wisteria sinensis*) which has always been grown here, the family held dances and parties.

ABOVE: The Botanic Garden provides walking routes leading though a variety of flora – including orange orchards, palms and desert plants.

OPPOSITE: The Garden has always featured sculptures – some on permanent display, others brought in for special exhibitions.

3.3m/10.8ft around its trunk. Wisteria (*Wisteria sinensis*) dating from 1857 is still trained along the Lorings' arbour: there are twelve specimens that now snake up into the surrounding trees, and in May, the whole area is filled with its lilac flowers and their wonderful scent. There are waterfalls, streams, pools and aquatic plants, under canopies of trees such as araucarias (*Araucaria heterophylla* and *Araucaria bidwillii*) and huge magnolias underplanted with giant white birds of paradise (*Strelitzia nicolai*). At the top of the historic Garden there is a forest route

and a viewpoint route dating from the 1920s, both with native plants. Another area near the *Mirador* has a superb collection of cacti and succulents, and as you leave the Garden you walk through an avenue of prehistoric plants.

The map you're given on arrival says you are very likely to get lost in the Garden...but advises you not to worry, as this is one of the place's charms. This is undeniably true, and on your visit you will encounter an exceptional collection of plants and trees from around the globe.

Giardini Botanici Hanbury
La Mortola, Italy (1867)

The stunning 18-hectare (44-acre) *Giardini Botanici Hanbury*, also known as La Mortola, lies a few kilometres from the resort town of Ventimiglia, northern Italy, surrounded by the Mediterranean Sea on the *Riviera dei Fiori*. Just a skip and a hop from France and Monaco, this creation of two English brothers, Thomas and Daniel Hanbury, has earned a reputation as a world-class centre of acclimatization research. Today, it's operated by the University of Genoa, with students of horticulture arriving from across the world and staying in the guest house in the Gardens.

The History
The Hanbury family were the Quaker owners of the London chemists, Allen & Hanbury (later known as Allen & Hanburys), and became rich through the sale of blackcurrant pastilles. Daniel, the elder brother, was a botanist and pharmacologist who also studied pharmacognosy (the study of medicines derived from plants). His brother Thomas, meanwhile, set himself up in Shanghai, China, as a textiles and property trader, earning a fortune by exporting cotton to the USA during the American Civil War. The two brothers were to make a garden in Italy, one using his great botanical knowledge and the other financing it through his immense wealth.

It was Daniel who found the perfect site while holidaying on the Riviera. In 1867, the now retired Thomas returned to Europe and bought the land. Although not a botanist, he was interested in creating a subtropical garden where he could experiment with introducing plants from different subtropical environments.

A Local Matter
The brothers employed local workers in the Gardens, training a generation of gardeners who were to continue the tradition of floriculture in this part of the Italian Riviera. The Hanburys supported the building of local schools and an old people's home, endowing plantings and funding an entire department at the University of Genoa. When the novelist Arnold Bennett lunched at La Mortola in 1904, he described Thomas as 'the Lord God of these parts', recalling that he 'has the finest private garden in the world, 100 acres, 5,000 species (some absolutely unique) and 46 gardeners.' Two years earlier, Thomas had purchased the site of Wisley in Surrey, England, gifting it to the Royal Horticultural Society in 1903.

The German Ludwig Winter was appointed head gardener in 1868, having previously worked in the *Jardins des Tuileries* in Paris (from where he was sacked for singing the republican anthem 'La Marseillaise' in front of the Empress Eugénie). In his seven years at La Mortola, he established the Gardens' design, their plantings and systems of work.

Many of their exotic trees date from this period. The Hanburys acquired 47 types of exotic mimosa, along with aloes and agaves from South America and South Africa, and eucalyptuses from Australia (although most of the Gardens' eucalyptuses were lost during the Second World War). It was Daniel, who died in 1875, who planted the Italian

umbrella trees (*Pinus pinea*) along the edge of the Gardens. He also went plant-hunting for indigenous species growing in the wild, and introduced the cistuses – *Cistus albidus* and *Cistus salviifolius* – into the Gardens.

A Network of Botanists

The Gardens thrived in large part thanks to Thomas Hanbury's willingness to exchange knowledge and plants with other scientific institutions, gardens and individuals. To facilitate easier exchange, he drew up lists of cultivated plants and harvested seeds (the idea of such an *Index Seminum* having been originated in 1800 by Casimiro Gómez Ortega at the Royal Botanical Garden of Madrid). Kew Gardens became a contact, and Hanbury nurtured an important and enduring relationship with French botanist Gustave Thuret, who had a garden nearby in Antibes. Since 2017, La Mortola has twinned with Villa Thuret on scientific research, and today, La Mortola has links with 450 other botanical gardens.

A Question of Scale
The exact size of the Gardens has been the subject of some conjecture. Thomas Hanbury asserted that the original area was 40 hectares (98.8 acres), while the curator in 1912 claimed 45 hectares (111 acres). However, the present incumbent gives it as 18 hectares (44.4 acres), of which 10 hectares (24.7 acres) are cultivated gardens and 8 hectares (19.7 acres) natural woodland.

ABOVE: The bright red flowers of *Russelia equisetiformis* or the firecracker plant, a native of Mexico.

OPPOSITE: The Tempietto was brought from Kingston Maurward, another Hanbury property in England, and placed in the Gardens in 1947.

ABOVE: Sitting on the south-east corner of the Palazzo terrace is the Pavilion, designed by Thomas Hanbury, with its wonderful views over the Cape Mortola peninsula.

OPPOSITE: In amongst the rocks at the Hanbury Gardens is the *Agave attenuata* or the foxtail agave, whose flowers can reach up to 1.5m (nearly 5 feet) in height.

The Citrus

Thomas Hanbury's network allowed him to stock citrus cultivars otherwise unknown in western Europe, such as mandarins (later grown commercially) and the citron 'Buddha's finger' (*Citrus medica* var. *sarcodactylis*), acquired in Shanghai in 1880. By 1890, the Gardens had over twenty varieties. The collection today includes the Australian Round Lime (*Microcitrus australis*), probably the oldest *microcitrus* in Europe, and examples of pomelo (*Citrus maxima*) and bitter orange (*C. aurantium*), among many others.

A Family Affair

When Thomas Hanbury died in 1907, 7,000 people attended his funeral as a sign of respect.

His daughter-in-law, Lady Dorothy Hanbury, took over the Gardens, revitalizing them after 1918. They were badly damaged during the Second World War, and the family attempted to save and replant what they could. They remained in Lady Hanbury's hands until 1960, when she sold them to the Italian State. The University of Genoa took over stewardship in 1983.

Thomas Hanbury's daughter, Hilda, was also a plant collector, and she had introduced the *Caltha polypetala* into La Mortola from the Vatican Gardens in 1902. This giant Caucasian kingcup/giant marsh marigold was not grown elsewhere in Europe, and was closely guarded by the Swiss Guard, so how she managed to acquire it remains a mystery.

Penang Botanical Gardens
Malaysia (1884)

The Gardens sit at the bottom of Penang hills, in a valley described as 'an amphitheatre of hills' in the midst of tropical rainforests, and with the famous waterfall as their nearest neighbour. A great escape from the polluted roads, the Gardens are known as the 'green lungs' of George Town, and, due to their location, are also known as the 'Waterfall Botanic Gardens.'

As you arrive you are welcomed by the main gate of Penang Botanical Gardens with fountains and four large pools of aquatic plants, containing such gems as giant waterlilies (*Victoria amazonica*) and other remarkable lotus blossoms. Once inside, the layout is still that of Charles Curtis's formal English garden (see below), but with the added local fauna. The monkeys here are happy to steal your belongings. Since 2004, when the Penang Botanical Gardens were expanded into the surrounding forest (increasing their size to 242 hectares/598 acres of protected area), there has been a chance to see an even wider range of Malaysian animals, such as macaques, giant squirrels, and dusky leaf monkeys.

The History

The Botanical Gardens were founded in 1884 as a colonial nursery garden, the first in the Straits Settlements. There were two previous botanic gardens on Penang Island before this one; the first was set up by the East India Company in 1794 as a spice garden. Christopher Smith (d.1806), a botanist trained at Kew Gardens, was put in charge: he was also sent off to collect nutmeg and cloves in the Molucca Islands, to be planted in the garden. In less than ten years Smith had swelled the collection to '19,000 nutmeg and 6,250 clove trees…[in] a collection of some 33,000 'spiceplants''. After Smith's death in 1806 the garden and its plants were sold off by the island's Lieutenant Governor Robert T. Farquhar. In 1822, a second short-lived garden was founded on the advice of Sir Stamford Raffles; it was sold in 1834 by Governor Murchison, who apparently had no interest in gardens or botany. The location for both gardens was in the Ayer Itam valley, but the exact sites are unknown.

In 1884, Nathaniel Cantley (1847–1888), superintendent of the Singapore Botanic Gardens, arrived to establish a third garden on the present site, bringing in English botanist Charles Curtis (1853–1928). Recommended by Kew Gardens, Curtis had been Assistant Superintendent of Forests and Gardens in the Penang District under the Straits Settlements, a colony that included the trading ports of Penang, Malacca and Singapore. Curtis knew the area well, as he had been plant-hunting in the Dutch colonies for James Veitch & Sons' Royal Exotic Nursery at Chelsea, London. He was placed in charge of the Penang region of the Forests and Gardens Department, including the Botanical Gardens, as well as 3,575 hectares (8,834 acres) of forest reserves. The Penang Gardens were to be

OPPOSITE: A cascade at the Penang Botanical Gardens. Originally, the site was known as the Waterfall Botanic Gardens.

important for the dissemination of seeds and specimens throughout Asia and into England via Kew Gardens.

Curtis was also responsible for the design of the landscape; he reused a granite quarry and also incorporated old nutmeg plantation. His focus was on creating a beautiful English-style garden, with large open lawns and a surrounding carriage circuit from which the Gardens could be viewed. Some of his successors note that he was more interested in producing elegant, striking vistas than in the systematic or scientific arrangement of the botanic beds. Although Curtis was working in the colonial interest, specifically focusing on economically valuable plants, he had a personal fascination with Penang's native flora, which he introduced into the Gardens.

During the Second World War, the Botanic Garden became the Japanese

ABOVE: A flowering Cannonball Tree (*Couroupita guianensis*) in Penang Botanical Gardens. The tree is indigenous to the tropical forests of Central and South America.

RIGHT: Topiary trees are pruned into intriguing shapes and underplanted with colourful bedding.

occupiers' base on Penang: surviving in the Gardens are the remains of the tunnels and train tracks the Japanese used to transport their supplies; it is possible to see some of these by the lily pond.

Rare Plants

The ten rare cannonball trees (*Couroupita guianensis*), were planted in the Gardens in 1886, at the time of the British rule. The trees were brought in from Brazil, possibly to acclimatize and assess them.

ABOVE: Flowering *Bauhinia kockiana* is not only a colourful plant, but is used as a traditional medicine for treating skin conditions.

RIGHT: The Botanical Gardens have formal walks which lead on to jungle trails in the surrounding hills; these are of varying levels of difficulty.

BELOW: Many animals can be seen in the Gardens, including various species of monkeys.

They can grow to a height of 20m (65ft), and produce the strange orbed fruits that are the size and shape of cannonballs. The trees are known for their pleasant scent, which is given off to aid pollination by attracting insects.

Research

The Gardens have retained much of Curtis' design and have flourished as a landscape park. They are now under the control of the Penang State Government and aim to focus on conservational and educational programmes.

New York Botanical Garden
New York City, USA (1891)

The New York Botanical Garden is a serious research institution with a herbarium and library, but it is also a landscape garden, an arboretum, a living museum, and a vibrant exhibition venue with world-class exhibitions that blend art and horticulture in a new and exciting way.

In winter, the beautiful circular Enid A Haupt glasshouse is a warm place to discover thousands of plants from around the globe, from the palms of the tropical rainforest to the ferns and mosses of the cloud forest. In the courtyard of the Conservatory every autumn there is a colourful display of Japanese maples and chrysanthemums, grown in fabulous shapes to celebrate *Momijigari* (the Japanese tradition of commemorating autumn leaves). In springtime the Orchid Show takes place, each year with a different theme – and at the same moment in another corner of the Garden the wonderful collection of lilac trees bursts into life. Their display was first designed in the 1940s by Marian Cruger Coffin, one of the earliest women landscape architects in America. In the summer months, the Rose Garden created by another great woman designer, Beatrix Farrand, with its thousands of scented and colourful blooms, must also be seen.

The History

America's primary centre for the study of plants before the Revolutionary War had been Philadelphia, but by the early 1800s, New York City was becoming pre-eminent in plant biology scholarship and learning. Medical practice still involved extensive herbal knowledge, and New York's College of Physicians and Surgeons and the city's Columbia College both knew they had to address the need for botanical studies.

Dr David Hosack (1769–1835), who taught botany at Columbia College, realized that his students required a botanic garden in order to learn from living plants. He was aware that European universities had possessed such facilities since the 1500s, and in 1801 he founded the first one in America: the Elgin Botanic Garden, on the site of today's Rockefeller Center, but then several miles outside the city. Hosack invested his own money in it, made Order Beds, built an elegant glass conservatory, and wrote a catalogue of its plant collection. But he struggled with the garden's finances, and within a decade it had closed down.

European Inspiration

The New York Botanical Garden as we know it today was founded by a husband and wife: Dr Nathaniel Lord Britton (1859–1934), a Columbia University professor of botany and geology, and Elizabeth Gertrude Knight Britton (1858–1934), a renowned scholar of mosses. In 1888, the Brittons had travelled to London on their honeymoon and visited the Royal Botanic Gardens at Kew. Inspired by what they saw, they returned to New York with the intention

RIGHT: The New York Botanical Garden's Enid A Haupt Conservatory, built in 1902, was inspired by mid-19th century London glasshouses like the one at Kew Gardens.

LEFT: The dome of the Enid A Haupt Conservatory has been restored and upgraded technologically, and is now computerized.

OPPOSITE: The Beatrix Farrand-designed Peggy Rockefeller Rose Garden is a riot of scent and colour in the summer.

of establishing a similar institution in America. Their campaign to achieve this coincided with the efforts of newly rich industrialists of the Gilded Age – J Pierpoint Morgan, Andrew Carnegie, John D Rockefeller and Cornelius Vanderbilt II – to finance the structure of the city. These so-called Robber Barons were putting New York firmly on a level with the great European cities they sought to emulate. They were founding institutions such as the American Museum of Natural History (1869), the Metropolitan Museum of Art (1870), the New York Zoological Society (1895) and the New York Public Library (also 1895). The city's Botanical Garden was to be another part of this cultural infrastructure: it was set up in 1891 with Cornelius Vanderbilt as the first president of its board, and in 1896, Britton was made its first Director.

Britton went in search of a suitable location for the Garden, and city officials offered him the 101-hectare (250-acre) Bronx Park as a possible site. It was perfect, with rolling hills and a freshwater river in a rock-cut gorge, as well as a forest of 20 hectares (50 acres). Calvert Vaux (1824–1895), one of the designers

of Central Park (along with Frederick Law Olmsted), produced the first layout for it: this was approved in October 1895, and although Vaux died the following month, the landscaped Garden was ready for public opening by 1900.

Britton also introduced a series of botanical explorations that continue today, with studies being conducted in South America, notably in the rainforests of the Atlantic coast of Brazil and the foothills of the Andes Mountains.

The Peggy Rockefeller Rose Garden

In 1916 acclaimed landscape architect Beatrix Farrand (1872–1959) was asked to collaborate on a design for the rose garden with the Horticultural Society of New York. Farrand was to create an elegant design out of a difficult triangular site, with a central iron gazebo, and latticed fencing all around it. However, shortages of materials during the First World War meant that the garden was only partially built – and it remained incomplete until the 1980s, when the gazebo and fencing were finally erected due to the patronage of David Rockefeller, who named the garden after his wife. Restored in 2007, the Rose Garden now contains over 750 varieties of both historic and modern roses.

Innovative Exhibitions

The Garden has more than a million visitors annually, with crowds continually drawn in by the innovative exhibitions held on its site. In recent years there have been some incredible shows, including 'Brazilian Modern: The Living Art of Burle Marx' which was a large-scale botanical tribute to the Brazilian modernist artist. In 2021, the gardens exhibited spectacular sculptures and immersive installations by influential Japanese artist Yayoi Kusama. The exhibition, titled 'Kusama: Cosmic Nature', celebrated the artist's profound connection with the natural world, inspired by the patterns, colours and life cycles of flowers and plant life.

Educational Programmes

The Garden participates in educational programs that are larger and more diverse than those of any other garden in the world. Currently, 100 Ph.D.-level scientists are engaged in 25 international

ABOVE: Yayoi Kusama's 'Kusama: Cosmic Nature' exhibition at the NYBG took place in 2021, where she debuted her 'Dancing Pumpkin' sculpture of 2020.

OPPOSITE: Another exhibit in 'Kusama: Cosmic Nature' was the 'I want to Fly to the Universe' sculpture which was inspiringly placed so that it reflected in the water of a pond.

collaborations in 49 countries. In 2013, The New York Botanical Garden launched its Humanities Institute, to support interdisciplinary research between the arts and sciences. The Institute brings scholars to the LuEsther T Mertz Library to research relationships between humanity and nature, landscapes, and the built environment. The William and Lynda Steere Herbarium opened in 1901 and now holds 7.5 million dried plants. The NYBG is currently creating an online, virtual version with digital images that can be sourced by scholars across the world.

The National Botanic Gardens of Ireland
Dublin, Ireland (1895)

Ireland's National Botanic Gardens are situated in Glasnevin, only 3km (2 miles) from the centre of Dublin. A second site at Kilmacurragh in Wicklow is the former estate of the Acton family that came under the stewardship of the National Botanic Gardens in 1996 (although a longstanding association was established through Glasnevin's Victorian-era curator, David Moore). The Gardens were founded with commercial and utilitarian considerations in mind but were soon growing plants that promoted what botanist John Lindley was to term 'systematic botany.' This remit widened out even more when exotic plants started to arrive from foreign shores, the original agricultural purpose took a back seat to botanical investigations, and Dublin started to link up with scientists in other gardens, particularly Kew and Edinburgh.

History

In the Middle Ages, when there were no Irish universities with medical schools, scholars returned from studying in Montpellier and began translating medical texts into Irish – including *De Materia Medica* (the 'Encyclopaedia of Dioscorides'), a pharmacopeia of medicinal plants dating from the era of the Emperor Nero. However, it was only in 1790 that the Irish Parliament, supported by the Speaker of the House, John Foster, provided funds to the Dublin Society (now the Royal Dublin Society) for a public botanic garden. It was a further five years before a suitable site was found. Rather than having the exclusively medical focus of many other botanic gardens, it was established with a view to promoting a more scientific approach to the study of agriculture.

The Gardens first opened to the general public in 1800, and over the next forty years the collection grew to include a large number of specimens unrelated to agriculture but considered beautiful or interesting in themselves. This expansion was facilitated by the arrival of plants from around the world via importers such as Messrs Veitch, and by closer contact with the great gardens in Britain, notably Kew and Edinburgh.

In 1834, the Glasgow-born Ninian Niven took over as curator. He oversaw a comprehensive redevelopment of the Gardens' layout, introducing many of the features present today. He published a *Visitor's Companion* and visitor numbers leapt from 7,000 a year to over 20,000 during his tenure. By the time Niven resigned in 1838, he had paved the way for a 'golden age' at Glasnevin under the directorship of David Moore.

David Moore

Moore, like Niven, was a Scotsman, born in Dundee in 1808. After migrating to Ireland as a twenty-year-old, he became a foreman at Trinity College's Botanic Gardens in Dublin and undertook landmark surveys of County Antrim and Londonderry.

After taking over at Glasnevin, he travelled widely throughout Europe expanding its collections. He also incorporated specimens from Australia, where his brother Charles was director of the Sydney Botanic Garden. David

ABOVE: The Teak House, one of the early glasshouses in Dublin's National Botanic Gardens. This is the location for various exhibitions, including the bonsai and orchid shows.

proved successful in orchid cultivation, supervising the country's first orchid germinations from seed in the late 1840s.

In order to cultivate rare species unlikely to thrive at Glasnevin, he used his good relations with various estate owners. These included the Actons, whose estate at Kilmacurragh is now the Gardens' second permanent site. In 1878, Glasnevin transferred from the jurisdiction of the Royal Dublin Society to the government and is today administered by the Office of Public Works. When Moore died in 1879, his 22-year-old son Frederick succeeded him and consolidated the Gardens' burgeoning international reputation.

The Glasshouses

As more and more tropical plants arrived in Dublin it became crucial for the Gardens to build glasshouses. The most exciting and beautiful of them was the Curvilinear Range, designed by one of the most important glasshouse architects of the 19th century, local ironmaster Richard Turner. The Glasnevin range had a new style of curvilinear glass which not only looked beautiful but allowed more light to pour in. Opened in 1849 by Queen Victoria, today the range – the most important buildings in the National Botanic Gardens – have been thankfully restored. They are divided into two wings, with a central dome: the east wing contains plants from the mountains of Southeast Asia, while the west wing includes Australian, South African and South American floras.

The Victoria Waterlily House

Five years after Victoria cut the ribbon on Turner's masterpiece, the Duncan Ferguson-designed Victoria Waterlily House, financed by a fête held in the gardens in June 1853, opened its doors. Like many of its contemporaries (see, for instance, Leiden), Dublin felt the need to build a home for the spectacular *Victoria amazonica*, the giant Amazon waterlily with leaves up to two metres in diameter, spiky underneath and smooth on the surface. Initial efforts at growing the waterlilies were unsuccessful, but in 1855 the Oxford University Botanic Garden sent seeds to Ireland which flowered at the first attempt. Now awaiting much-needed restoration, it is to be hoped that

ABOVE: The Curvilinear glasshouses at Dublin's National Botanic Gardens date from 1843 to 1869 and were built by Richard Turner.

OPPOSITE: This glasshouse, whose architect was Sir Charles Lanyon, dates from the 1840s; its dome, a tall central section for palms to grow into, was added in 1852.

Potato Famine

David Moore was the Gardens' director at the time of the Irish potato famine (1845-52). His observations of the impact of disease on potatoes within the Gardens, made on 20 August 1845 after having heard reports of the blight, represent the first documented record of the disease in the country. He correctly identified the fungus (*Phytophthora infestans*) causing the blight – although, despite his best efforts, he was unable to find a cure for it.

the House will soon be full of waterlilies again.

The glasshouse complex is completed by a Teak House, Alpine House, Cactus and Succulent House, Great Palm House and Orchid House, between them covering plants from environments ranging from the frozen tundra to the tropics.

Herbarium

Glasnevin is also home to the National Herbarium, comprising a collection of over half a million dried and documented plant specimens from Ireland and the rest of the world. Its economic collection alone contains around 20,000 specimens of

ABOVE: The large pond brings much biodiversity to the garden, acting as a magnet for wildlife such as frogs, toads, insects and birds, and creating homes and food for all manner of creatures.

OPPOSITE: The Garden stretches over 20 hectares (50 acres), with more than 17,000 plants which come from all areas of the world.

fruits, fibres, plant extracts, seeds, wood, fibres and other plant artefacts. Along with an associated library, the Herbarium is an internationally recognized reference centre and repository for the study of Irish and international botany.

Jardín Botánico Carlos Thays
Buenos Aires, Argentina (1898)

This triangular garden is enclosed by busy roads, and yet inside is an unexpected oasis of calmness, right in the centre of Buenos Aires. The land was originally chosen for its proximity to the major parks in the city, although the heavy road traffic couldn't have been anticipated over a hundred years ago. The Garden has a total area of 7 hectares (17 acres), and contains approximately 6,000 species of plants, trees and shrubs, as well as five greenhouses, and a number of monuments and sculptures.

History

The *Jardín Botánico* was designed by the French-Argentine landscape architect Carlos Thays (1849–1934) in 1898. He used three distinct 19th-century garden styles, Formal, Gardenesque and Picturesque, to create visual interest, and wanted the Garden to be used by the people of Buenos Aires for pleasure, and not just as a site for the study of science and botany.

Thays had left France to design the *Sarmiento* public park in the Argentinian city of Córdoba. He was the star pupil and assistant of Édouard François André (1840–1911), the *Jardinier Principal* of Paris, who had been one of the team reshaping that city under Georges-Eugène Haussmann. Thays moved to Buenos Aires to take up the post of Director of Parks & Walkways in 1891, and was to design most of its squares and public spaces. He planted over 1.2 million trees there, focusing on native species. André's influence can be seen in Thays' planning of the tree-lined boulevards and public gardens, which gave the city a French feel.

Thays and his family lived in a red-brick, English-style mansion in the *Jardín Botánico* between 1892 and 1898. The building, which dates from 1881, is still in use, and currently serves as an art gallery and exhibition hall. The city government's gardening school also operates from there, and the site includes an important botanical library and a herbarium which are accessible to the public.

French, Roman and Japanese Gardens

There are small examples of three different styles of garden in the *Jardín Botánico*. Next to the mansion where Thays lived is the Roman Garden with a statue of Mercury at its centre. It has a selection of plants, including topiary trees, cypresses, box, roses, acanthus, and laurels, that the first-century Roman, Pliny the Younger wrote of having in his garden. Thays wanted to illustrate the most widespread garden styles in the history of landscape. The French Garden is modelled after the 17th- and 18th-century formal Baroque style, inspired by the work of André Le Nôtre at Versailles with flower parterres with scrolling box. The Oriental Garden holds plants from Asia, including the first *Ginkgo biloba* planted in Argentina.

RIGHT: The so-called 'Decorative Fountain' at the *Jardín Botánico*; there are no surviving records about its origins. It is planted up with papyrus and waterlilies.

ABOVE: The bright yellow flowers of *Peltophorum dubium*, known as *birá-pitá* in Argentina. The tree is also planted along the wide avenues of Buenos Aires.

OPPOSITE: Built in the *Jardín* in 1881, this English-style mansion house, once lived in by Carlos Thays, is now used as an exhibition space.

The Garden Cats

Pet cats have been so consistently abandoned in the Garden – at a rate of one per day in summer – that a group of volunteers have set up the *Asociación Civil Gatos Botánico* to feed and vaccinate the animals, provide them with veterinary care, and try to find new homes for them. Several hundred now live on the site, and – thankfully – seem happy.

The Glasshouses

Perhaps the most delightful element in the *Jardín* are its five glasshouses. The oldest, largest and most beautiful is a unique *Art Nouveau* steel and glass structure. It was brought to Argentina in sections shipped from the Paris Exhibition of 1899, where it had been on display. The glasshouse, thought to be the only one of its kind, holds many species of tropical plants, with nearly a thousand specimens of ferns, orchids and palm trees. In April 1996 it was declared a National Historic Landmark.

Plant Collections

The plants and trees here are grouped into regions, specializing in native Argentine flora, such as the Tipa tree (*Tipuana tipu*), the cedar tree (*cedro salteño*) and the yellow flowering *Peltophorum dubium*, known as *ibirá-puitá* in Argentina. There are more than five thousand species of plants and trees from all over the world, from Asia to Africa to America.

Sculptures

The grounds also feature a large collection of sculptures, and a monument called the 'Weather Indicator', designed by José Marcovich and donated by the local Austro-Hungarian community in 1910. The inclusion in the Garden of a 1909 bronze copy of the challenging *Saturnalia* is somewhat surprising. The original, a group sculpture by sculptor Ernesto Biondi dating from 1899, is on display at the *Galleria Nazionale d'Arte Moderna e Contemporanea* in Rome. It was shown in the *Exposition Universelle* (1900) in Paris, where it won the Grand Prix. *Saturnalia* became controversial when it was removed from display at the Metropolitan Museum of Art in New York in 1905: the trustees there were said to have been appalled by its 'immorality.'

OPPOSITE: *Saturnalia* – bronze sculpture by Ernesto Biondi. It was placed here in 1909; four years earlier, its original had shocked the trustees of New York's Metropolitan Museum of Art.

BELOW: There are over thirty pieces of sculpture dotted around the *Jardín*. Some are copies of classical pieces; others are associated with people connected to the site.

Botanischer Garten und Botanisches Museum
Berlin, Germany (1910)

The Berlin-Dahlem Botanic Garden, comprising an area of just over 50 hectares (126 acres), is one of the world's largest and most important botanical gardens, and is now part of the Free University (*Freie Universität*) of Berlin. The Botanical Museum (*Botanisches Museum*), the Herbarium, and a large scientific library and archive are attached to the Garden.

In the 15.7 hectares (39 acres) of the plant-geography section you can travel all the way around the different plants of the Northern Hemisphere. The twelve rock gardens display plants from a diverse range of climates, from the Alpine and Himalayan mountain ranges to the forests of East Asia and the flower meadows of the great North American prairies.

The History

In the 16th century there was a historical ancestor to the present Botanic Garden of Berlin-Dahlem – a kitchen and herb garden in the grounds of Berlin's City Palace, said to have been made by gardener Desiderius Corbianus. In 1679 these were converted into model agricultural gardens, laid out in the city's Schöneberg district, on the orders of the Great Elector, Frederick William of Brandenburg. Two hundred years later, the Garden was relocated to Dahlem, due to the lack of space in Schöneberg.

The present Garden was constructed between 1897 and 1910 under the directorship of Adolf Engler (1844–1930), and designed by him as a landscape park. Its original use was as a holding place for colonial plant material, exotics being

brought back from the German colonies. It is composed of three sections: the Arboretum (14 hectares/24.5 acres), a collection of trees, woody plants and roses; an open-air area which is organized by the geographical origin of the plants (13 hectares/32 acres); and a third area, dedicated to systematic botany, and containing around 1,500 plant species. After the outbreak of the First World War in 1914, its staff numbers were radically reduced, and the arrival of plants from the colonies soon stopped. No plant-collecting trips were possible during the war years or the immediate post-war period. Expeditions began again with the one to Chile and Bolivia led by German botanist Erich Werdermann from 1923–1927. Werdermann (1892–1959) had become the curator of mushrooms in the Herbarium of the museum in 1921, and he later became curator of the Garden.

A 1938 inventory known as the *Dahlemer Handliste* counted around 18,000 species of plants. The Garden was badly damaged in the air raids of 1943 during the Second World War, although in 1947 it was discovered that parts of the Herbarium and Library had survived in the Russian sector of Berlin. During the post-war rebuilding of the Garden, a decision was made to emphasize the holdings of wild origin species and to concentrate on the

RIGHT: The *Botanischer Garten und Botanisches Museum* of Berlin is the second-largest botanic garden in the world, with only Kew Gardens in London beating it in size.

verification of their taxonomic identity. Presently, about half of the 32,500 cultivated accessions and 20,750 different species in the Garden have been verified.

Glasshouses

There are sixteen glasshouses, and the spectacular Gothic tropical house is, at 60m (196ft) long and 23m (75ft) high, one of the largest in the world. It is a tropical paradise containing pools and waterfalls, and its tall trees are covered with epiphytes (air plants), all underplanted with colourful plants. Here you will find the unusual *welwitschia*, endemic to the Namibian desert: it has only two leaves but an enormous underground stem. The glasshouses, including the glass Cactus Pavilion and the Victoria Pavilion, have

ABOVE: Among the glasshouses here is one of the largest glass structures of its kind in the world.

OPPOSITE: This glasshouse boasts a wide diversity of plants; there are tangles of vines and the sharp blades of tropical and subtropical leaves.

a total area of 6,000 square metres (64,583 square feet). The Victoria Pavilion holds a collection of orchids, carnivorous plants, and the giant white waterlily *Victoria amazonica*.

The Medicinal Garden

You will find this area between the Herbaceous Plants and the Useful Plants beds. An interesting newcomer to the Garden, the Medicinal Garden

is designed in the shape of the human body, with individual plants placed on the part of it they are used for. It contains over 230 medicinal plants, and shows their pharmacological importance. The plants' labels provide information on their applications, active substances, conservation aspects, and potential dangers or toxicity.

Herbarium Berolinense

The Herbarium of the Berlin-Dahlem Botanic Garden and Botanical Museum was founded in 1819. The destruction of the scientific library and most of the Herbarium on 1 March 1943 was a very sad event in the history of the institution. However, it is estimated that about half a million specimens were saved, and among this remaining stock there are at least single specimens from nearly all the old collections. After the war, Erich Werdermann ordered the reconstruction of the Garden and greenhouses and the remaking of the library and Herbarium. Werdermann took overall charge in 1955;

he died four years later, and is buried at the Garden, next to Adolf Engler.

The Herbarium is now the largest in Germany, holding a collection of more than 3.8 million preserved specimens. All plant groups – flowering plants, ferns, mosses, liverworts, and algae, as well as fungi and lichens – are represented in it. Associated with the general Herbarium are special collections of dried fruits and seeds, wood samples, and specimens preserved in alcohol.

The Seed Exchange

The Botanic Garden's seedbank or genebank has been storing seeds from native wild plants since 1994, and is one of the oldest in Germany.

The Garden publishes a yearly seed list with up to 2,000–3,000 items; these are available for exchange with other institutions for scientific purposes. As with its live holdings, the major emphasis is on seed collected in the wild, or from garden plants of known wild origin.

OPPOSITE: The Garden was planned between 1897 and 1910 by Adolf Engler and architect Alfred Koerner, and boasts wide strolling pathways.

BELOW: A more formal area in the Garden, revealing remainders of its 19th-century design layout, with topiary and pools.

Brooklyn Botanic Garden
New York City, USA (1911)

Once a boggy marshland, Brooklyn Botanic Garden (BBG) is now a beautiful, verdant oasis amongst the bustling cityscape of New York City. Today, the garden is regarded as one of the very best displays of urban horticulture: it aims to connect people to the world of plants and encourage us all to take care of the environment.

The History
Historically, the site the Garden now occupies was once part of the lands belonging to the indigenous Lenape people. The Garden recognizes this and, with the Lenape Centre in Manhattan, is working to address and raise awareness of the culture and acknowledge the Lenape territory.

In the 19th century, after American landscape architect Frederick Law Olmsted designed Prospect Park, the City of Brooklyn decided to purchase additional acres of land in 1897 with the hope of this additional park space becoming a garden. An unprepossessing site, it was mainly the remains of an ash dump. It took until 1910 for the Garden to become reality, and the Olmsted Brothers (sons of Frederick Law Sr), Frederick Jr and John Charles Olmsted, were given the job of designing and laying out the original site plan. The garden was then developed over the next thirty years by landscape architect Harold Caparn, who took charge of the project in 1912.

The Brooklyn Botanic Garden's first director was American botanist Charles Stuart Gager, and he was to serve for thirty years in the position, establishing the site as a centre of education. In 1917, the Laboratory Building and Conservatory (now Administration Building and Palm House) was created by the firm of McKim, Mead & White, architects of many of the USA's Gilded Age mansions.

Two early gardens
In 1911, when the Brooklyn Botanic Garden officially opened, the first display garden was the Native Flora Garden designed by BBG's first Curator of Plants, Norman Taylor (1883–1967). He was a botanist (previously based at the New York Botanical Garden) with a special interest in endemic plants, this led him to walk almost 2,000 miles whilst studying the flora of Long Island. This garden only contains plants that grow within a 100-mile radius of the city of New York City.

The Japanese Garden
The Japanese Hill-and-Pond Garden was completed in 1915 by Japanese-American landscape designer Takeo Shiota (1881–1943), and it was one of the first public Japanese gardens in the United States. Shiota combined two distinct Japanese styles, the ancient hill-and-pond design and the traditional stroll-garden of the 16th-century Azuchi–Momoyama period.

RIGHT: The Japanese Hill-and-Pond Garden by Japanese-American landscape designer Takeo Shiota was one of the first public Japanese gardens in the United States.

ABOVE: The Desert Pavilion, with its distinctive tinted windows, displays cacti and other plants from two hemispheres.

Brooklyn Botanic Garden's Children's Garden

In 1914 a Children's Garden was established with the aim of allowing children to grow flowers, vegetables and herbs, and learn about nature. Kids as young as six can begin to develop basic botany and horticulture skills, and even younger toddlers can become budding gardeners. Over a thousand children a year take part in growing crops with the assistance of garden instructors – instilling an interest in community horticulture, sustainability and an understanding of conservation.

The Desert Pavilion

This pavilion has collections of cacti and other succulent wildflowers from the world's arid desert regions. The eastern side of the house has plants from North and South America, and the western side of the house holds plants from across Africa, the Middle East and Australia. The Desert Pavilion was designed to show the evolutionary difference between the plants from the two hemispheres, and this also allows for year-round interest. As one group of plants settles down for winter, the other group is gearing up for summer, allowing visitors to always see plants in bloom. Heating is controlled by a simple venting system for climate control, otherwise the plants are dependent on the hours of sunlight that Brooklyn gives them.

New York Metropolitan Flora project

For hundreds of years botanists have been collecting information on the local flora of New York and its surroundings and disseminating their findings. This work is being continued in the Garden's New York Metropolitan Flora project (NYMF), which was founded by Steven Clemants in 1990. The project aims to record all the flora within a 50-mile radius of New York City, in an attempt to understand the urban landscape. The research takes in all of Long Island and includes south-eastern New York State, northern New Jersey and Fairfield County, Connecticut, a total of approximately 19,813 sq km/7,650 square miles.

BELOW: The Cherry Esplanade is not to be missed for the *Hanami* in springtime.

Cherry Esplanade

One of the prettiest areas in the Garden is the Cherry Esplanade, which from late March to early May is a riot of blossoming trees in all shades of pinks and creams. Its broad green lawn is bordered by avenues of twenty-six species and cultivars of cherry: these create a pink petal carpet as the flowers start to drop. The cherries are all carefully chosen; they include Prunus 'Hata-zakura' – this cherry tree has a soft pink blossom that becomes white when fully open. Prunus Hata-zakura descends from an original cherry tree that grew in the Hakusan Shrine in Tokyo.

The month-long cherry blossom season, called *Hanami*, is a Japanese tradition of flower-viewing which is

celebrated at the Garden. The Garden's website has information on timings for the moments of flowering, along with cultivar descriptions and locations, plus an online map and guide. Surrounding the cherries are avenues of scarlet oak trees (*Quercus coccinea*), the Liberty Oaks, which were planted to commemorate the victims of 11 September 2001.

Bluebell Wood

In late April, head just south of Cherry Esplanade to find a magical sea of vibrant blue. More than 45,000 bluebells (*Hyacinthoides hispanica* 'Excelsior') are planted here in the woodlands of oak, birch, and beech trees, and for just a couple of weeks of the year they put on a magnificent display.

Herb Garden

With an orchard, medicinal dye and textile beds, this lovely garden is also an

ABOVE: The stunning, newly designed Robert W. Wilson Overlook, with its slopes and ramps and planting for all the seasons.

RIGHT: The ecologically sustainable Diane H & Joseph S Steinberg Visitor Center, with its 891 square metre (9,600 sq ft) living roof.

educational resource. It is wonderfully designed with many joyful touches – huge cabbages grow in the Knot Garden, and the whole is enclosed with espaliered fruit trees.

Rock Garden

Just to the north of the Herb Garden is the Rock Garden, created in 1916 with boulders from the last Ice Age that had been unearthed on-site and then arranged to create miniature habitats for alpine plants. It was the first feature of its kind in an American public garden.

Kirstenbosch National Botanical Garden
Cape Town, South Africa (1913)

Kirstenbosch is one of the world's great botanic gardens, and one of the largest. Together with its spectacular neighbour, Table Mountain National Park, it makes up part of the Cape Floristic Region Protected Area, which became a UNESCO World Heritage Site in 2004. It is a wonderful Garden, although only around 6% of its territory is cultivated, and the rest is a protected area of natural forest and fynbos.

Its collection of more than 7,000 species is dominated by plants indigenous to southern Africa, with some 2,500 from the Cape specifically. Operated by the South African National Biodiversity Institute (SANBI), it is one of the country's ten national botanic gardens. Famous for its Protea Garden and a Cycad Amphitheatre, it also has a Fragrance Garden, a Medicinal Garden – and, in 2002, a Useful Plants Garden was added, focusing on African and indigenous uses of plants.

Today, the Garden's primary concern is to 'champion the exploration, conservation, sustainable use, appreciation and enjoyment of South Africa's exceptionally rich biodiversity for all South Africans.'

The History
The history of the Garden is bound up in the troubling and difficult story of colonialism and apartheid. When the first European visitors arrived from Portugal in the late 1400s, the Khoikhoi people were living and working on the land, and had been doing so for about 2,000 years. Later, when the area was occupied by settlers from the Dutch East India Company, they planted a hedge of native wild almond trees (*Brabejum stellatifolium*) to keep out the Khoikhoi: part of it still exists. For many in South Africa, this hedge is seen as one of the first beginnings of apartheid, with the exclusion of indigenous people from their land by white settlers. From the 17th century onwards, there were violent clashes between the Dutch and later the British using the land for agriculture with enslaved people doing the labour.

Like so many botanic gardens of the 19th century, the Garden here was steeped in colonialism and the economic botany and acclimatization of plants that would finance the British Empire. The taking of land and plants from the indigenous peoples to found the Botanical Garden in 1913 needs to be addressed.

Even the former approach to the labelling of botanic specimens is revealing. Popular names in English or Afrikaans were included, but no connections were made to the African and indigenous names by which the plants were known. These are areas that the Garden is beginning to put right, with labels that, in addition to Latin taxonomy, also carry names in English, Afrikaans and African languages as standard.

RIGHT: Kirstenbosch must have one of the most glorious locations of any botanical garden, nestling under Table Mountain in Cape Town.

Birth of a Garden

In 1903, Henry Harold Pearson, a Cambridge University botanist, arrived in the colony from England to become the first Professor of Botany at the South African College (now the University of Cape Town).

In 1913, the Cape government designated the Kirstenbosch area for a botanic garden, but provided a tiny budget of just £1,000 per year. The Botanical Society was formed in the same year, in the hope of creating public interest and support. Henry Harold Pearson took on the role of the Garden's unsalaried director, and J W Mathews was appointed its first curator. The Dell was the focus of their early cultivation, the area becoming a sea of cycads.

Pearson died of pneumonia in 1916 and was buried at Kirstenbosch, where his contribution is recognized in an epitaph: 'If ye seek his monument, look around.'

Further development was curtailed by the First World War, but under Pearson's successor, Robert Harold Compton (who arrived in 1919) and Mathews, many of the Garden's most popular features were created in the years that followed. These included the Cycad Amphitheatre, Mathews' Rockery, the Arboretum (now with a 130-metre (427-foot) treetop walkway arcing above the canopy), the Koppie (a rocky outcrop with specimens of the Pelargonium and Geranium families), and the spectacular Protea Garden.

The Herbariums

The Compton Herbarium was named in honour of the second director of Kirstenbosch, Professor R H Compton, who founded it in 1937. In 1956, the country's oldest herbarium, the South African Museum Herbarium, was transferred to Kirstenbosch. It had been established in 1825, with 325 specimens

OPPOSITE: The Centenary Tree Canopy Walkway is a curved steel and timber bridge that skims the tops of the trees in the Arboretum.

BELOW: The glasshouse of 1998 contains plants could not survive outside in the local climate, as well as those that require special growing conditions.

Colonel Bird's Bath

Colonel Bird's Bath is a bird-shaped pool surrounded by tree ferns, African holly, emerald fern and bush and fireball lilies, and it significantly predates the Garden itself. Fed by four sparkling fresh-water springs, it was built from Batavian brick by Colonel Christopher Bird, the Deputy Colonial Secretary, while he briefly owned Kirstenbosch in 1811.

donated by the Danish botanist Christian Friedrich Ecklon, and was significantly developed later in the century by the Cape's first colonial botanist, Dr Karl Wilhelm Ludwig Pappe. The Stellenbosch Government Herbarium, set up by Dr Augusta Vera Duthie in 1902, became the next herbarium to transfer to Kirstenbosch in 1996. Encompassing some three-quarters of a million dried specimens – including amaryllis, heaths, orchids and proteas – the Herbarium is in a modern research complex.

The Round Hut

Built in 2003 in the Useful Plants Garden, the Round Hut symbolizes the decolonization of the Botanical Garden at Kirstenbosch. It was developed in a communal way, with the assistance of master hut builder, uTata Zanazo, and designed in a wattle and daub style used by the amaMpondo people in the Eastern Cape.

The Dell

Today, the Dell is a glorious shady haven of tall trees, where ferns and assorted shade-loving species like *Streptocarpus*, Impatiens and *Mackaya bella* flourish. Originally, it was packed with oak trees, but the density of their roots and leaf-fall made it virtually impossible to cultivate much else. Though the oaks stayed for many years due to popular demand, they are gradually being replaced with more appropriate local species, like *Anthocleista grandiflora* (forest fever tree) and assorted yellowwoods.

The large Botanical Society Conservatory exhibits specimens from an array of different environments, from savannah to fynbos and karoo. The Garden is home to many bird species, and although much of the fauna is nocturnal, it is possible to spot mongooses, otters, tortoises and terrapins, antelopes, genets and porcupines.

OPPOSITE: The Kirstenbosch region has two main vegetation types – fynbos and forest. Both can be explored in the Garden by taking various mapped trails and hikes.

BELOW: A male Southern Double-Collared Sunbird photographed in Kirstenbosch National Botanical Garden in Cape Town. He is one of its many wonderful avian visitors.

Huntington Botanical Gardens
California, USA (1919)

Located in San Marino, California, the Huntington Botanical Gardens' 52.6 hectares (130 acres) are home to over 83,000 living plants, including many rare and endangered species, across sixteen themed gardens. Part of a world-class complex of botanic, academic and cultural interest, they include a laboratory for botanical conservation and research.

The History

'The Huntington' was founded in 1903 as an educational, research and cultural centre by Henry Edwards Huntington (1850–1927), a railroad magnate. Purchasing the working San Marino Ranch south of Pasadena, with its 243 hectares (600 acres) of established agricultural land including citrus groves, nut and fruit orchards, and alfalfa crops, he had transformed it, by 1919, into what was to become the world-famous Huntington Library, Art Museum and Botanical Gardens.

Huntington was a collector of plants, art and books, and many of his purchases can still be seen throughout the Gardens, such as the fountains and statues he bought from Italy. Experts were brought in to assist with the development of the estate. Architects Myron Hunt and Elmer Grey designed the house; Joseph Duveen advised and chose the art; and landscape architect William Hertrich (1878–1966) was instrumental in developing the various plant collections that comprise the foundation of The Huntington's botanical gardens.

Gardens on a Theme

The sixteen themed gardens include the Australian Garden, on a plot that was once an orange grove, replanted in the 1940s with a thousand eucalyptus trees from the US Department of Agriculture to test their viability as timber. These trees were drastically thinned in the early 1960s, with only the most interesting and attractive specimens retained. The survivors were then interplanted with informal groups of native Australian plants. Eucalyptus (*Myrtaceae*) are perhaps the plants most associated with Australia as they are exclusive to the country except for a few species to be found in New Guinea and the Philippines. There are in total over 700 species, growing in virtually every habitat type except for rainforest and some desert regions. At The Huntington, you can see around a hundred Australia eucalyptus species, including the endangered *Eucalyptus woodwardii*, a small tree or mallee with beautiful lemon flowers endemic to Western Australia (and known as Gungurra by the aboriginal Noongar).

Early spring is the best time to visit this garden. Start with the acacias in all sizes, with their scented yellow blooms, and continue through the blooms of the fuzzy flowered kangaroo paws (*Anigozanthos*), various melaleucas, wax flowers, and vivid blue hibiscus.

OPPOSITE: The Helen and Peter Bing Children's Garden with its artichokes and spiralling topiary.

The *Frances and Sidney Brody California Garden*, spanning 2.6 hectares (6.5 acres) and landscaped with nearly 50,000 California natives and dry-climate plants, reflects the local Mediterranean climate as well as the history of the ranch. Meanwhile, the Camellia Garden echoes a particular interest of William Hertrich, who began the collection. Only one specimen predates him, the oldest camellia in the Garden, a 'Pink Perfection.' Among the other garden themes are Desert, Herb, Japanese (in which a reconstruction of a 320-year-old Japanese magistrate's house was opened in 2023), Jungle, Lily, Rose and Shakespeare.

The Garden of Flowing Fragrance or *Liu Fang Yuan*

A recent addition to The Huntington is this 21st-century interpretation of a classical Chinese garden – designed by Jin Chen and Jim Folsom, and widely considered to be the finest example outside China. The masterplan for its 6 hectares (15 acres) was made in Suzhou, eastern China, and then adapted to the modern needs of American visitors in California.

This is a particularly apt setting, as The Huntington is home to an exceptional collection of Asian art and literature, and there is a large Chinese-American community in Los Angeles. Cleverly using the principles of landscape design as found in Suzhou during the Classical Garden period of the 16th and 17th centuries and combining them with the Garden's actual location, this is a garden of international collaboration.

BELOW: The Rose Hills Foundation Conservatory for Botanical Science has three different areas: lowland tropical rain forest, a cloud forest, and a carnivorous plant bog.

The Huntington's *Amorphophallus titanum*

In the summer of 1999, the Gardens became the focus of attention when their *Amorphophallus titanum* ('corpse plant', Titan arum) flowered for the first time in Southern California. In just two weeks, over 100,000 people queued to witness the inflorescence. The same plant bloomed again in 2002, a surprise as it was thought at the time that the plant bloomed just once and died. Now they have forty-three 'corpse' flowers in their collection, and many are the offspring of the 2002 bloom. This bloom began The Huntington's initiative to propagate and pollinate and send a plant to all the botanic gardens in the USA.

ABOVE: The Huntington Desert Garden is over 100 years old, and has one of the largest collections of succulents and cacti in the world.

Some elements of it were built in China and the work on the site was done by both Chinese and American landscape architects, builders and gardeners. The masterplan for the garden was made in Suzhou and then adapted to the modern needs of the American visitors back in California.

The plants have been chosen for their symbolic meanings and literary or

BELOW: The Huntington's Chinese Garden – 'Liu Fang Yuan' or 'the Garden of Flowing Fragrance'.

RIGHT: 'The Lake of Reflected Fragrance' in the Chinese Garden. Years of research and planning went into the Garden's creation; it draws its inspiration from the gardens of Suzhou.

cultural significance, while respecting the Garden's Californian location. There is a bamboo grove and a recently planted Chinese medicinal herb garden. At the heart of the Garden is the Lake of Reflected Fragrance, which in Chinese symbolism represents the *yin* (the calm), surrounded by scholars' rocks (brought from Lake Tai in China) that represent the *yang* (the stability). There are ten pavilions, a grotto, and viewing platforms linked by pathways, each with many layers of meanings and literary inscriptions and carvings to reflect upon. Botanical Director of the Gardens Jim Folsom, one of its creators, says: "What ultimately makes this Garden special is what you bring to it or do with it. When visitors come to the gardens, they create their own meanings."

Botaniska Trädgård
Gothenburg, Sweden (1923)

Stora Änggården, an idyllic 19th-century country estate with rolling meadows, surrounded by forests, was to become the nature reserve Änggårdsbergen – planted up with exotic trees to form an arboretum. Today this land belongs to the Botanical Garden of Gothenburg. With its nature reserve it makes up one of the largest botanic gardens in the world at 175 hectares (432 acres),and boasts fauna such as hare, fox, deer and elk, and extraordinary flora in its glasshouses and outdoors.

The History

The city of Gothenburg was founded in 1621, and to mark its 300th anniversary, it was decided by the City Council to create a Botanical Garden (*Botaniska*) and nature park. The Garden was to be used for experimentation and biological demonstrations, as well as for education and recreation. By 1919 the first part of the site, the Woodlands, had opened to the public, followed by the *Botaniska* in 1923.

The botanist and explorer Carl Skottsberg (1880 – 1963), who had studied at Uppsala University and later become conservator of its Botanical Museum, was in charge of creating the new *Botaniska* from 1915. In 1919 he was appointed Professor and Director of the Garden. From the beginning, education and scientific research have been essential aspects of its purpose, and since 1936, the *Botaniska* has actively collaborated with the University of Gothenburg in this work.

The Climate

The Garden has to contend with the long cold, snowy season from November to March, when the average daily high temperature is below 5.5°C. The coldest month of the year in Gothenburg is February, with an average low of -3.3°C and high of 1.5°C. The summers are more comfortable, but often cloudy, and this climate limits what can be grown in the outside areas of the Garden, although they still manage to have 16,000 species planted in the open air, and 4,000 species in the greenhouses.

The Japanese Glade

During the 1920s, when the Garden was still being developed, a seed exchange with Sapporo Botanic Garden (*Hokkaido Daigaku Shokubutsuen*) began. This connection with and interest in Japanese species continued, and in 1952, a seed-collecting expedition to Japan took place. The Gothenburg Garden's Director, Professor Bertil Lindquist (a forest geneticist), and botanist and dendrochronologist Tor Nitzelius, brought back many new discoveries from the wild habitats in East Asia. An area in the Garden was designed to house the plants, and in 1952 the Japanese Glade was founded. The idea was not to recreate a Japanese garden but to reference Japanese culture and nature.

OPPOSITE: The original Glasshouse. New glasshouses are currently being designed for the Garden in collaboration with the London-based Harris Bugg Studio and consulting architects COBE.

ABOVE: The flower beds at the Gothenburg botanical gardens treat visitors to a riot of colour.

OPPOSITE: The waterfall in the Rock Garden. Thousands of different plant species thrive in this spectacular habitat.

The following year, the Arboretum was also established, featuring an Asiatic area within which a number of Japanese trees were planted.

The Rock Garden
A recent addition to the Garden in 2012, the Rock Garden has as its focus a stunning, cascading waterfall. Displayed here are a collection of native Scandinavian plants now threatened by extinction, arranged according to their geographical origin. This area of the Garden can be visited from April to September.

The Herb Garden
The Herb Garden has as its centrepiece an 18th-century summerhouse, designed by architect Bengt Wilhelm Carlberg for his country estate, Kärralund. Among the many plants found here are ones that have had a long history of use by the native people of Scandinavia such as the *Sámi*, including angelica (*Angelica archangelica*), which grows wild in the mountains and on beaches along the Swedish coast, and has been cultivated in Scandinavia since the Vikings. The *Sámi* use angelica to prevent infections, not only in its natural state but also preserved in reindeer milk.

The Glasshouses: Rare plants and abundant orchids

The glasshouses hold the collection of 1,500 orchids which are the finest anywhere in Sweden.

There is also an incredibly rare flowering Toromiro tree (*Sophora toromiro*), which is extinct in its natural habitat on Rapa Nui (Easter Island), but thriving here in the glasshouse. The Botanical Garden is part of an international group working to conserve the genetic diversity of Easter Island's wild and cultivated vegetation. This project includes endeavouring to reintroduce the Toromiro tree to Easter Island. So far they have been unsuccessful in their attempts but the group is determined to continue and hopes for future success.

Some of the Garden's special collections include the Dionysia Family, which are small cushion-like plants of the *primula* genus that have adapted to difficult habitats. In the wild, they live on vertical cliff faces and in the mouths of caves.

Geophytes: the bulbs and tubers collection

The Garden holds the world's largest collection of botanical bulbs and tuberous plants, again brought here straight from their wild habitats. When Norwegian botanist Per Wendelbo became Director in the 1960s he began the collection of botanical bulbs. His botanical interest was mainly on regions rich in geophytes, and his research focused on countries in the Middle East, namely Turkey, Iran and Afghanistan. Wendelbo was a plant collector, and material he gathered in the wild was later planted in the Garden. His work collecting in the field has been continued by present staff members, who carry on the expeditions and add to the Garden's holdings.

OPPOSITE: The Palm House of 1878 stands in the centre of the Garden. It was modelled on the Crystal Palace in London.

ABOVE: Seasonal planting at Gothenburg is always impressive. Here, Allium 'Globemaster' is seen in June.

Jardin Majorelle
Marrakech, Morocco (1923)

Marrakech is known as the 'rose-hued city': full of palm trees (over 180,000 of them), it is an oasis in the desert. On *Avenue Yacoub El Mansour*, a street full of merchants selling fruit from their carts, is a garden, atelier and house created by an artist in the 1920s, and later saved and revived by one of the world's most talented fashion designers. This is the *Jardin Marjorelle*: and entering it transports you from the chaos of the outside world and a busy desert city into a botanic garden full of mystery and magic.

The History
Developed from a 1.6-hectare (4-acre) palm grove, planted with poplars, originally called *Bou Saf Saf* ('the poplars' in Arabic), the Garden was created over a 40-year period by the French Orientalist artist Jacques Majorelle (1886–1962). From 1923 it was his home, studio and workshop. As he extended the land he began to build, bringing in the French architect Paul Sinoir to design a villa and studio in a Moorish Art Deco style. Majorelle transformed the palm grove into a botanic garden, introducing plants from around the world. He was still living and working at the site when he opened the Garden to the public in 1947. It stretches to nearly a hectare (2.4 acres) today, with a plant collection from five continents and 135 plant species.

The buildings are painted in Majorelle's trademarked colour, *Bleu Majorelle*, which he said was like using a jeweller's black velvet to display gemstones. The colour was inspired by the shade of blue used in Moroccan tiles, and the traditional headdresses of the Tuareg people. *Bleu Majorelle* forms the backdrop for the botanical collection, and the pots and plants are framed against it. In contrast, the paths are paved in red-dyed concrete, while the colourful pots were introduced by the garden's later owner, the iconic French couturier Yves Saint Laurent.

Saving the Garden and the Restoration
After Majorelle's divorce, he sold the Garden in the 1950s, and it fell into a state of neglect. In 1980, Saint Laurent and his partner, the French industrialist and collector Pierre Bergé, were living next door to it. When its widowed owner died, news circulated of plans to destroy it and replace it with a hotel complex, and Saint Laurent (a long-time admirer of the Garden – in his own words, 'the *Jardin Majorelle* has provided me with an endless source of inspiration, and I have often dreamt of its unique colours') stepped in to save it. He and Bergé purchased the property, and began restoring it. With the assistance of American architect Bill Willis and French decorator Jacques Grange, they also worked on the house, renaming it *Villa Oasis* (after a book by the French socialist author, Eugène Dabit) and adding a café/restaurant, bookshop and gift shop.

RIGHT: *Jardin Majorelle,* Marrakech – a view of the square pool and fountain in front of the studio building. They were painted in this vibrant blue during the 1930s.

ABOVE: The Garden's collection of succulents and cacti was the first to be made in Morocco.

OPPOSITE: The long water rill – again painted in *Bleu Majorelle,* the colour trademarked by artist Jacques Majorelle – stretches along the Garden.

Musée Berbère

Majorelle's blue (of course) Art Deco studio now houses the *Musée Berbère* (or *Musée Pierre Bergé des Arts Berbères*, to give it its full name), a museum exhibiting some 600 pieces dedicated to the culture and traditions of the Berber people of North Africa. It includes a brilliant mirrored chamber displaying a collection of jewels and enamelled objects.

American landscape designer Madison Cox was brought in to work on the private areas of the Garden, eventually becoming the general director there. Beginning work in the late 1990s, he introduced a number of new plants and assisted in the restoration of the existing pavilions, various architectural details and some paths. Conscious of climate change, he removed the Garden's parched lawns and replaced them with gravel.

Yves Saint Laurent died in 2008 and his ashes were scattered in the Garden. After Bergé's death, Madison Cox was appointed president of both the *Fondation Jardin Majorelle* and the *Fondation Pierre Bergé - Yves Saint*

Laurent, which has museums in Paris and Marrakech. Cox says his intention is to preserve the Garden, and he has brought in Marc Jeanson, a brilliant young botanist, to exhibit and identify the plants.

The Plants

The Garden's plants were mostly collected by Jacques Majorelle over many years. Some he brought from local regions, and others from his frequent travels further afield. They are classified into five categories (cacti, palms, bamboo, blooming potted plants, and aquatic plants), but different species are mixed together. The plants create shady spots where visitors can escape the intense Moroccan heat, with a walk through the bamboo groves providing peace, calm and coolness. You can see the traditional use of the water rill and the pergola with its hanging and trailing vines, and both features introduce a lushness.

The plants are labelled with their scientific names, along with illustrations

ABOVE: The square pool and fountain are connected to the long rill which runs the length of the Garden.

RIGHT: In summer, the pools of the *Majorelle* garden are covered in waterlilies. Around its edges are pots of specimen plants under the shady canopy provided by the palm trees.

for easy recognition. The flora tempts in an array of exotic birds, with well over sixty bird species having been recorded as visiting the Garden.

The *Jardin Majorelle* has become one of the most visited sites in Marrakech, and there are some 900,000 sightseers a year. To accommodate this huge number, it was expanded in December 2018 by opening up the section containing *Villa Oasis*, where Bergé lived until his death in 2017. Along a pathway dripping with bougainvillea, this separate garden displays giant succulents, cacti and mature palms, alongside water features filled with fish, noisy frogs and lush lily pads.

Kyoto Shokubutsuen
Kitayama, Japan (1924)

Japan has more than sixty botanic gardens and a great love of plant life. The Kyoto Prefectural Botanical Garden is the oldest botanical garden in Japan, and it holds over 120,000 plants from 12,000 species. It is conveniently situated along the scenic Kamo River, in Kitayama. The Garden is spectacular whatever the season: in spring the world-famous cherry trees blossom; in summer the fragrance and colour of the roses are prominent; in autumn the rich reds and orange of the maple trees can be seen; and for the rest of the year there are an exciting variety of trees, flowers, lush open spaces and even a forest to discover.

The Garden covers 24 hectares (59 acres), and this land is divided up into various areas. There is a Perennial Garden, a Peony Garden, an annual Bonsai Exhibition, Bamboo, Camellia and European Style Gardens, and many others. In the northern half, there is the only natural forest in the Garden, the *Nakaragi-no-mori*, and a botanical ecological garden, where native plants growing wild in the mountains and fields of Japan are planted in a state that is as close to nature as possible.

The History

The *Kyoto Shokubutsuen* was established on land donated by the wealthy Mitsui family of bankers and financiers to the City of Kyoto. Its creation was a tribute to Japan's Emperor Taishō, whose reign had begun in 1912, and it opened to the public on 1 January 1924.

Following the Second World War, the Garden was requisitioned for twelve years in order to accommodate the families of the military forces stationed in the country. After this, it started to undergo restoration, and was reopened to the public in 1961.

Sakura

In Japan, the ancient tradition of enjoying the blossoms of cherry trees is called *hanami*, which means 'flower viewing.' The Garden is famous for the wide variety of cherry trees: there are about 500, and they blossom in a riot of shades of pink. This is the place to come for cherry blossom *hanami* without the crowds. The garden has 200 types of Sakura or Japanese cherry (*Prunus* genus), including rare species, and because of the wide variety of trees the season is longer here than usual, from mid-March to late April. The Garden also celebrates with a cherry blossom festival and a night light-up event during the season. The Sakura Light-Up is usually held at Kyoto Botanical Garden from late March to early April.

Hydrangeas and Lotus

During the rainy season in June, the Hydrangea Garden bursts into full bloom. It has over 189 types of Hydrangeas and over 2,500 plants. Many Japanese shrines and temples also have gardens featuring Hydrangeas, and they hold

OPPOSITE: Tulips and blossoming Sakura (Japanese cherry) trees in the *Kyoto Shokubutsuen*. Spring is one of the best times to visit the Garden, when '*hanami*' or flower viewing takes place.

special openings and events such as the *Ajisai Matsuri* (Hydrangea Festival) during the peak season where visitors can enjoy the flowers, as well as special foods, ceremonies, and light-ups after dark. Next to the Hydrangea Garden is the pool of Lotus: throughout July and August, when the Lotus are in full bloom, it is extraordinary to walk across the boardwalks into the centre of the pool and view the site from there.

The Conservatory Complex

The unusual conservatory stems from it being designed to resemble the nearby Kitayama Mountains, and its interior rooms recall the nearby *Kinkaku-ji* Temple. Designed in 1992 and built with an iron frame, it contains over 4,500 taxa divided into environmental areas: the climates range from temperate to wet tropics and dry tropics. As you wander

ABOVE: *Echium wildpretii*, common name 'Tower of Jewels', is an architectural plant that can grow to 2 metres (over 6 feet) in height. It provides nectar and pollen to bees and other insects.

OPPOSITE: A spinning *mizuguruma* (Japanese water wheel) blends in perfectly with the rock pool here.

along the pathway, the various collections are revealed – these include tropical bromeliads, desert and savannah plants, succulents, orchids and temperate alpines.

A highlight of the dry tropics area in the conservatory is the *Adansonia digitata*, common names for which are the African baobab, monkey-bread tree, upside-down tree, cream of tartar tree and the Tree of Life. It is native to Africa and the southern Arabian Peninsula, where it has been used for centuries for its medicinal values. The tree is deciduous, losing its leaves for eight months, coinciding with the dry season; and it is perfectly adapted to drought conditions as it stores water in its trunk It flowers in both the wet and dry season with huge white flower heads that hang downwards from the branches. The blooms open in the afternoon and last for only one night, in the wild they are pollinated by bats and bushbabies.

The 'Useful Tropical Greenhouse' is devoted to the tropical plants that have been used incessantly for colonial exploitation, such as cocoa (*Theobroma cacao*), coffee (*Coffea*), banana (*Musa*) and mango (*Mangifera indica*).

Centenary

The Kyoto Botanical Garden celebrates its 100th anniversary in 2024, and continues to improve its 'living plant museum', developing education and research.

Night Garden

Among the unusual plants that can be seen in Kyoto's Night Garden is *Epiphyllum oxypetalum* – a night-blooming cactus,

native to Southern Mexico and areas of South America but naturalized in China, and commonly known as 'princess of the night' or 'queen of the night.' 'Epiphytic' flora take their name from Greek words, the prefix 'epi-' ('upon') and 'phyton' ('plant'): they attach themselves to other plants, share their sources of nutrition and water, but aren't parasites. Other inhabitants of the Night Garden with similar characteristics include the critically endangered (CE) *Rafflesia arnoldii* or corpse lily. It has no leaves or stem – just a vile-smelling flower – and is found in the jungles of Borneo and Sumatra.

The Ginger Plants

On the southwest side of the path connecting the large lawn and the viewing greenhouse is an area filled with ginger lilies which can be found blooming in Autumn. A cultivar of the genus *Hedychium* of the Zingiberaceae (ginger) family, many of the species have edible, decorative or medicinal properties. The flowers vary in colour from orange and red to white, and are fragrant; they have been used by the Japanese as decorations since the Edo period. Kaichi Oyama (1906–1989) of the Kagawa Prefecture created a range of ginger lily varieties, of which thirteen types such as Kinkaku, Momo no Teru, and Genpei can be seen in the botanical garden.

RIGHT: Another beautiful time to visit Kyoto's *Shokubutsuen* is in autumn, when the trees in the Garden put on an incredible show as the leaves turn to red, gold and orange.

Jardin Botanique de Montréal
Québec, Canada (1936)

Located in the heart of Montréal is one of the world's largest and most beautiful botanic gardens. It occupies 75 hectares (185 acres) of vibrant green space, with arboretums, living sculptures, ten exhibition greenhouses and thirty thematic gardens. It was designated a National Historic Site of Canada in 2007, and is regarded as one of the most important botanic gardens in the world due to the extent of its rare collections and resources.

The History

The Garden dates back to the time of Canadian botanist and scientist Conrad Kirouac, known as Brother Marie-Victorin (of the Brothers of the Christian Schools) (1885–1944). An avid botanist, his book *La Flore Laurentienne* (1935) is still considered a major reference on indigenous plant life in southern Québec. From 1919 he initiated plans to found a botanic garden in Montréal, but these were to take some years to come to realization. Finally, in 1935, he made a passionate plea to the Mayor for the creation of a botanic garden:

We will soon be celebrating Montréal's three hundredth anniversary. You need to give a gift, a royal gift, to the City, our city. But Montréal is Ville-Marie, a woman... And you certainly can't give her a storm sewer or a police station... It's obvious what you must do! Give her a corsage for her lapel. Fill her arms to overflowing with all the roses and lilies of the field!

Brother Marie-Victorin chose landscape architect, horticulturist and botanist Henry Teuscher (1891–1984) as his designer, and in 1936 Teuscher was officially appointed the new Garden's Superintendent and Chief Horticulturalist. He was ideally equipped for the roles, having held various senior positions at the Botanical Gardens in Berlin and New York. The garden landscape Teuscher designed was inspired by the British picturesque style, and within it he included space for both scientific and educational functions.

Display areas

There are more than 20,000 species of plants under cultivation here, in some thirty themed areas.

The Rose Garden has over 10,000 rose specimens, some dating from the 1860s. There is a Conifer Arboretum, a Forest with 45 collections covering 40 hectares (99 acres), and there are garden pavilions inspired by Beaux-Arts and Art Deco styles. The Marie-Victorin Herbarium contains specimens representing 99% of all of the plants found in Québec.

The Arboretum

When Henry Teuscher first saw the site, it included a still-functioning municipal stone quarry, and he had the idea of making a hanging garden there. The first trees were planted at the entrance to the Garden and near the administrative building in 1936. At the same time, lilacs,

RIGHT: A goddess sculpture in plants, part of the Mosaïcultures Internationales held in Montréal's Botanic Garden in 2013.

alders, birches and willows were planted around the ponds. Towards the end of the 1940s, the first collections were established in the south of the current Arboretum: these included crab-apples, mountain ash, cherry, plum and pear trees. Finally, in the 1960s, the stone quarry was finally filled in, and Teuscher was to see the Botanic Garden expand its arboretum there.

Chinese Garden

The Chinese Garden is a demonstration of the close relationship between the Shanghai City Parks Department and the Montréal Botanic Garden. The thousands of pieces needed for its construction were shipped from Shanghai to Montréal and assembled by fifty Chinese craftsmen in 1990. Reflecting the ancient art of Chinese gardens, its four main elements – the architecture of the pavilions, the plants, water and stone – create harmony and balance. Its Dream Lake Garden of 1991 was designed by Japanese architect and landscape architect Le Weizhong,

who was at that time director of the Shanghai Institute of Landscape Design and Architecture.

Japanese Garden

The Japanese Garden consists of three distinct areas: a strolling garden, a dry garden and a tea garden. It covers an area of 2.5 hectares (6 acres), using a combination of stone, water and plants to create an idealized version of nature. Strolling gardens (*kaiyū-shiki*) originated in Japan's Edo period (1603–1868), and are intended to display a sequence of effects from a path which twists around them.

The Garden was designed in 1988 by Ken Nakajima (1914–2000), an internationally recognized Japanese landscape architect, who wanted to create a garden using the natural beauty and topography of the site. Its aims were to promote a better knowledge of Japanese culture and to foster, develop and maintain cultural and social exchanges with Japan.

OPPOSITE: The important Art Deco-style Administration building was designed by architect Lucien Kéroack.

BELOW: A collaboration between the Parks Department of the City of Shanghai and the Montréal Botanic Garden produced the Chinese Garden; the overall concept came from the landscape architect Le Weizhong.

Strange Plants

The Garden has a number of unusual inhabitants, and one of the oldest is *Beaucarnea recurvata*, commonly called the elephant's foot or ponytail palm: it arrived from Paris as seed in 1938. The plant is considered critically endangered; it is native to Mexico, but can now only be found in the country's state of Veracruz. It naturally grows in dry zones, and the base of its trunk (the *caudex*) is designed to store water, enabling the tree to survive for 7-8 months without rain.

Alpine Garden

The diversity of the flora of the mountains and their boreal regions is examined in the Alpine Garden, with plants from the arctic tundra and locations from the Rockies and Himalayas to the Alps. The Alpine Garden has the largest collection of living plants in the Botanic Garden, with a huge number of species. This section of the Garden was begun in 1937, but was not completed until 1962, due to the disruption of the Second World War and the complexity of the build. A 'mountain' was constructed by building up from a clinker foundation with limestone or stratified dolomite from Saint-François in Laval, Québec: these stratified rocks are millions of years old. Plants were chosen from Southern Québec and added to the Garden in 2001: they promote the conservation of rare and threatened species from that area.

Another section, added in 2002, was made under the direction of Josef Halda, a distinguished botanist from the Czech Republic, in co-operation with the

ABOVE: The waterfall and cascade in the Japanese Garden, which was designed by landscape artist Ken Nakajima (1914-2000) and opened in 1988.

RIGHT: The Stone Boat in the Chinese Garden. It is a replica of the larger Marble Boat in the Summer Palace in Beijing.

Société de plantes alpines et de rocaille du Québec (Québec Alpine and Rock Garden Society). This garden is situated to the right of the entrance to the Alpine Garden, in front of the pine trees. Four years later the Vertical Crevice Garden was designed by a second specialist from the Czech Republic, Zdeněk Zvolánek. It recreates the natural habitat of eroded cliffs, with its stones positioned vertically, creating the habitat to allow plants to thrive that otherwise would not.

Surrounding the Alpine Garden is a large collection of conifers, with magnificent specimens such as the ponderosa pine, just one of the remarkable trees in the Garden.

Fairchild Tropical Botanic Garden
Florida, USA (1938)

Just south of Miami, along Biscayne Bay, there is a huge, lush landscape with tropical gardens, rainforests and hummingbirds: this is Fairchild Tropical Botanic Garden, one of the foremost such places in the USA. With 3,400 species, including orchids, native and exotic plants, all arranged throughout its Uplands and Lowlands ecosystems and exhibits, the Botanic Garden is a scientific organization committed to research, education, and conservation.

The Fairchild's botanic collection is extensive and includes many plant species that are endangered in their native habitats, as well as palms, cycads, flowering trees and shrubs, vines, and fruit trees.

The History

In 1938, the Garden's benefactor, Colonel Robert H Montgomery, opened the 33.5-hectare (83-acre) institution he named for Dr. David Fairchild, a plant hunter and one of the great botanists of his age. Fairchild travelled to more than fifty countries, sending over 200,000 edible or useful plants back to America, and simultaneously founding the U.S. Department of Agriculture's Office of Foreign Seed and Plant Introduction.

The Design

The Fairchild was designed by the landscape architect William Lyman Phillips, a consultant for the Olmsted Brothers who went on to work on projects across the world including the Panama Canal Zone, Puerto Rico, France, Italy, Switzerland, and Germany. On his return from his travels he renewed his association with Frederick Olmsted Jnr. and began work on one of his most rewarding projects, the Fairchild Tropical Botanic Garden in Coral Gables. Phillips was to expand the original garden, with landscape architect Noel Chamberlin working on the initial master plan. As consultant for the collections, he became expert in the tropical plants of the region.

Phillips' design works by opening up the Garden with a wide central avenue,

ABOVE: The Fairchild Tropical Botanic Garden is home to many endangered and rare plants.

RIGHT: There are over 500 palm species at FTBG, plus a special 5.2 hectare (13-acre) site, the Montgomery Palmetum, dedicated to them.

like a grand French Baroque vista. As you walk along the main axis, you encounter pathways which send you down intimate, shaded routes and into garden rooms. In these areas, Phillips used taxonomic families as a way of organizing the garden – individual collections of plants designed as composed scenes.

Dividing the Garden into upland and lowland areas is a bare rock escarpment: the upper section alternates between grass lawns and pools, and the lower area meets in the Palm Glade. In this lower section, which is a marine salt flat, Phillips contrasted the densely planted formality with softly controlled open spaces and lakes drawn from natural sinkholes.

The planting in the Garden switches from the very formal plant arrangements near the Garden Club of America Amphitheatre, to the more naturalistic plantings near the Museum and Nell Montgomery Garden House (both buildings designed by Miami architect Robert Fitch Smith).

The Tropical Flowering Trees Collection

The Fairchild has over 740 species of flowering trees from all over the world. The most famous flowering tree here is the cannonball tree (*Couroupita guianensis*): a member of the Brazil nut family, it was planted in 1938. The flowers of the cannonball tree are arranged on long stalks which project from the trunk, quite unlike any other flower: they are large, beautiful, and sweetly fragranced. The tree gets its common name from its

fruit, which mimic the shape and size of cannonballs.

The Tropical Conservatory, historically known as The Rare Plant House

The Conservatory is a 1,533-square metre (16,500-square foot) glasshouse which held a collection of over 450 species of tropical and subtropical plants, but suffered from complete devastation when Hurricane Andrew hit the Garden in 1992. Many of the plants were lost, but the glasshouse was restored and renewed by generous donors who funded the complete renovation of its fabric and the reinstating of the plantings. The collection has since gone from strength to strength with the addition of an incredible sculpture in vibrant primary colours by American glass artist and sculptor Dale Chihuly, named *End of the Day Tower*. It was created in 2005 for the *Chihuly at Fairchild* exhibition and is now part of the Botanic Garden's permanent collection.

Water features

There are eleven lakes and seven pools, some in the outdoor landscape and others in the glasshouses used to display the aquatic plant collections. The Pool in the Mood Sunken Garden is surrounded by lush tropical planting, with a cascading waterfall as its centrepiece. In the Sibley

RIGHT: Glass artist Dale Chihuly has held three exhibitions at FTBG. He has said, 'It is one of my favourite places in the world to show my work.'

The Fairchild Tropical Botanic Garden has plenty of spots to sit and admire the lush marshland vistas.

Victoria pool area are a selection of water lilies, the most dramatic being the *Victoria cruziana* and the *Victoria* 'Longwood Hybrid' (a hybrid between two species of the *Victoria* genus *V. amazonica* and *V. cruziana*). In their native South America these lilies can grow leaves of up to 1 metre (3 feet) across.

Hidden away in a grove of palm trees is the Amphitheatre pool, home to a wonderful selection of plants, including several unusual varieties of water lily: among them are the lavender *Nymphaea* 'Key Largo', named after Key Largo in Florida, the purply-blue 'Star of Zanzibar', pops of colour from the Scarlet rosemallow (*Hibiscus coccineus*), and the deep greens of the giant horsetail (*Equisetum giganteum*).

Fairchild conservation work

In south Florida there is an endangered ecosystem of pine rocklands of which only small areas survive, and the Fairchild is playing a major role in their conservation and restoration. Many of the plants are on the U.S. government's rare and endangered species list, and the Fairchild has been working to study their ecological requirements, in the hope of relocating plants in the future. Loss of pine rocklands habitat affects biodiversity, and insects and butterflies are threatened with extinction due to habitation loss. At risk is the Florida Atala butterfly (*Eumaeus atala*), due to the over-collection of the cycad *Zamia pumila* (already endangered in the wild), which is the food plant needed by the Atala's caterpillar.

The Australian Botanic Garden
Mount Annan, Australia (1988)

A 20th-century botanic garden set up specifically for conservation purposes in 416 hectares (1,028 acres) of open space, Mount Annan is Australia's largest botanic garden. It's designed to take visitors on an exploration of the country's unique plant life; they can walk through Bushland, Cumberland Plain Woodland, and swathes of colourful native plants. The site has been described as 'a roofless museum', demonstrating the way people and the natural world interconnect. The natural woodlands and grasslands, with their endemic plants, are at risk if not constantly assessed and managed as they are here. Mount Annan is the Australian native plant garden of the Royal Botanic Gardens, Sydney, and it holds over 4,000 native and introduced plant species.

The History

As the Garden acknowledges, its site is on the traditional land of the Dharawal and Gundungurra people. In the 19th century this became part the Glenlee estate, owned by Scottish magistrate William Howe, who built a brick Georgian house here in 1824. In the 1850s Glenlee was bought by James Fitzpatrick and it remained in his family until 1979. The Royal Botanic Gardens took charge of it in 1984, and opened it to the public four years later. Its original name was Mount Annan Botanic Garden, and this was changed in 2011 to The Australian Botanic Garden, Mount Annan.

Rare and Endangered Species

The Garden is a refuge for rare and endangered species, and conducts seed collection and propagation. An example is its work with the Wollemi Pine (*Wollemia nobilis*, named after the discoverer David Noble), a species known only through fossil records until 1994, when living trees were discovered in Wollemi National Park in New South Wales. With fewer than a hundred specimens in the wild grove, the tree is classed as Critically Endangered (on the Red List of the IUCN – International Union for Conservation of Nature). Research undertaken at Mount Annan and the Royal Botanic Garden in Sydney showed that the plant could be cloned, and both gardens began to produce new trees with the intention of making Wollemi pine specimens available to botanical gardens around the world.

The water supply

The Australian Botanic Garden's main water supply is principally derived from a very special water canal built in the 1880s. This is in itself a feat of engineering: an aqueduct partly made of sandstone blocks thought to have been quarried from Mount Annan. Some stretches of the canal run underneath the Garden.

RIGHT: The Australian Botanic Garden specializes in growing native plants; on the edge of the flower beds is a grass tree, *Xanthorrhoea johnsonii*, named for L A S (Lawrie) Johnson (1926–1997), a former Director of the site.

Daniel Solander Library

Daniel Solander was a Swedish botanist who arrived in Australia with fellow botanist Sir Joseph Banks in 1779 aboard Captain Cook's HMS *Endeavour*. Solander and Banks recorded and collected thousands of species during their time in Australia. The Library founded in 1852 and named after Solander was Australia's first for botanical research: there are over 250,000 catalogued items, encompassing many forms of written and illustrated works and original paintings. The oldest book in the Library is a printed copy of Dioscorides' *De Materia Medica* dating from 1550; and very pertinent to the collection is Joseph Banks' *Florilegium* – a collection of copperplate engravings of plants collected by Banks and Daniel Solander between 1768 and 1771 on the *Endeavour* voyage.

A cutting-edge Herbarium

The National Herbarium of New South Wales comprises a collection of over one million plant specimens: it has been held at the Royal Botanic Garden Sydney for almost 170 years. Some of the earliest specimens were those collected during James Cook's *Endeavour* voyage of 1770 by Sir Joseph Banks and Daniel

LEFT: The Garden's website features a unique online calendar which alphabetically lists the plants that are commonly in flower during each month of the year. Each plant has a small location map so that it can easily be identified in the Garden.

Solander. The collection continues to grow annually and more space is required, so the whole collection is being moved to a specially designed facility at Mount Annan. This cutting-edge building is by the Australian architecture studio *Architectus*, and is inspired by the *waratah* seed pod of New South Wales's floral emblem. Inside there are six protected vaults that keep the collection safeguarded from all extreme weather events, pest and insect attacks. It is now possible to visit the Herbarium not only for research purposes but on public tours.

Research – Testing for plant vulnerabilities

'Living on the Edge' is a joint study being undertaken by the Garden and the Australian National University's Research School of Biology to try and understand how Australian plants cope with extreme temperature. It's looking at seventy species, and seeing the effects of thermal tolerance on them. In their introduction to the study, the researchers have observed that 'Although extreme temperature events are increasing in frequency and intensity, little is known about the thermal tolerance breadth of plants from extreme environments, where both heat and cold extremes are common. This information is crucial to understand species distribution and survival under future climate regimes.'

The Plant Clinic

The Plant Clinic is open to the public as well as industry, and has a wide remit – offering plant disease diagnostics as well as pathogen detection and plant DNA identification. The Garden understands the importance of comprehending and discovering plant diseases, which threaten biodiversity and have a devastating impact on native flora and fauna. The Garden hopes that the Plant Clinic's diagnostic services will be a crucial first step in plant disease management and therefore in plant health and conservation efforts.

RIGHT: The polished steel façade of the Australian Plant Bank, where plants and seeds are stored safely for the future. They will be used for conservation and research purposes.

ABOVE: The Plant Bank's architects, BVN Donovan Hill, ensured that their designs preserved and protected the natural scrubland and wild flowers encircling the building.

Tromsø Arctic-Alpine Botanical Garden
Norway (1994)

The world's northernmost botanic garden is to be found in Tromsø, northern Norway's largest city, on an island a little over 320 kilometres (200 miles) inside the Arctic Circle. Administered by the University of Tromsø and Norway's Arctic University Museum, it is unlike most other botanic gardens. You will not find, for instance, a glasshouse. Instead, it is a rocky landscape of boulders, waterfalls and drifts of plants. In winter, snow covers the Garden from October or November until April, and you can take your skis to experience its evergreen shrubs and snow-capped surroundings. From late November until mid-January the sun never rises, but any sense of desolation is compensated for by spectacular displays of the *Aurora Borealis* (Northern Lights). Most flowering takes place between May and October, and during the summer they have the midnight sun.

An extraordinary setting
The plants come from all continents, from arctic and alpine regions all over the world. The completely naturalistic settings replicate an array of habitats perfect for spectacular Arctic and Alpine species from cold climates around the world. Rare Arctic plants grow in the crevices between rocks and boulders, and, surprisingly, so do plants from Africa such as the white-eyed ice plant (*Delosperma basuticum*) from Lesotho. Plants are well labelled, something that is often overlooked in outdoor spaces of botanic gardens.

A Short History
The Garden is located on land that was previously an old farm known as Southern Breivika, renowned for its extraordinary views and its setting amid a birch forest. The site was donated in 1938 by Ms Hansine Hansen, the farm's last owner. It subsequently served as the home of Breivika High School and, from 1980, for parts of Tromsø's University. In 1994 the area was opened to the public, and the Garden can be visited all year round and twenty-four hours a day, with no gates and no admission fees.

The Collections
There are in total twenty-eight themed collections, specializing in Arctic and Antarctic plants, with an emphasis on traditional perennials and herbs from Norway, as well as species native to the Himalaya, South America and Africa. Many of them cannot be seen elsewhere as they need the coldest climate. For example, plants such as the *Micranthes mandchuriensis* (previously part of the Saxifrage genus) are notably difficult to cultivate in warmer places.

One of the newer structural additions is a large stone ridge in the south of the Garden, which replicates the environment of a moraine – the soil and rock left by a moving glacier. On its drier, warmer southern slope, it is possible to see plants

RIGHT: Tromsø's rock garden, with the rare, low-growing *Rhododendron lapponicum* or Lapland rosebay, found in barren, subarctic regions.

from traditionally warmer climes including Chile, Lebanon, South Africa and Turkey. On the north-facing slope, which collects the snow, there are species from Svalbard, a Norwegian archipelago in the Arctic Ocean, including one of Europe's rarest species, the *Ranunculus wilanderi*.

Primulas are a particular specialism: there is a huge collection with a number of very rare species amongst them, including some that come from an altitude of over 5,000m (16,404ft) in the Indian Himalaya.

From far-flung places

The Garden boasts an extensive collection of South American plants, many of which need a drier habitat found on steep, rocky slopes where sand and gravel are left exposed to the sun. Here you can see specimens of *Perezia recurvata* with its tiny star-shaped flowers, found in the wild in Argentina and Chile; and the bobbing red flowers of *Calandrinia caespitosa*, another native of Chile.

One of the most outstanding genus collections is that of the *Meconopsis* from the Himalaya which seem to love the climate here, including the brilliant blue, *Meconopsis grandis*, commonly known as the Himalayan blue poppy. The poppy is native to China, Tibet – and Bhutan, where it is the national flower. It is notoriously difficult to grow but has a long blooming season here.

The Rhododendron Valley holds new types of rhododendron that have been discovered at higher altitudes in the Himalaya. The Garden has the world's largest gentian collection, comprising 345 different species and a few cultivated hybrids. They grow in the northern hemisphere and are found in the high mountainous regions of North America, Asia and Europe – whose conditions the Garden here can replicate.

Plants from the gardens of North Norway

Curator and botanist Brunhild Mørkved has established a collection of plants sourced from the gardens of local people. Many of these are now considered weeds and are not cultivated elsewhere – species like larkspurs (*Delphinium*), monkshoods (*Aconitum*) and ranunculus, along with greater masterwort (*Astrantia major*) from central and eastern Europe (also found in the mountains of Norway) and tall meadow-sweets (*Filipendula ulmaria*).

OPPOSITE: As the winter snow thaws, the first flowers start to appear in early May, and the flowering season generally continues until October when the Arctic snow returns.

Index

Picture credits

References

Introduction and general sources

Londa Schiebinger and Claudia Swan (ed.), *Colonial Botany, Science, Commerce, and Politics in the Early Modern World*, University of Pennsylvania Press, 2005.

Gavin Hardy and Laurence Totelin, *Ancient Botany*, Routledge, 2016.

Lucile H Brockway, *Science and Colonial Expansion, the Role of the British Royal Botanic Gardens*, Yale University Press, 1979.

Arthur W Hill, "The History and Functions of Botanic Gardens." *Annals of the Missouri Botanical Garden*, vol.2, no.1/2, (1915): pp.185–240.

Clare Hickman, *The Doctor's Garden: Medicine, Science and Horticulture in Britain*, Yale University Press, 2021.

Lucia Tongiorgi Tomasi, "Projects for botanical and other gardens: A 16th-century manual." *The Journal of Garden History*, vol.3, no.1: pp.1-35.

D. Gandawijaja, S. Idris, R. Nasution, L P Nyman, and J. Arditt,"*Amorphophallus Titanum* Becc: A Historical Review and Some Recent Observations." In *Annals of Botany* 51, no.3 (1983): pp.269–278.

May Woods and Arete Warren, *Glass Houses*, Aurum, 1988.

India Hobson and Magnus Edmonson, *Glasshouse Greenhouse: Haarkon's world tour of amazing botanical spaces*, Pavilion, 2018.

Louise Wickham, "Botanic Gardens, Politics and Empire." *Gardens in History: A Political Perspective*, Oxbow Books, 2012: pp.187–210 .

Sathnam Sanghera, *Empireland, How Imperialism Has Shaped Modern Britain*, Penguin, 2021.

Edward W Said, *Culture and Imperialism*, Chatto & Windus, 1993.

Jill H Casid, *Sowing Empire, Landscape and Colonization*, University of Minnesota Press, 2005.

Maria Espírito-Santo, Maria & Ana Soares, & Manuela Veloso, *Botanic Gardens, People and Plants for a Sustainable World*, Isa/Leaf, 2020.

Nadine Monem (ed.), *Botanic Gardens, a living history*, Black Dog Press, 2007.

Sara Oldfield, *Botanic Gardens, Modern-Day Arks*, New Holland, 2010.

Christopher Woods, *Gardenlust, a botanical tour of the world's best new gardens*, Timber Press, 2018.

Esther Helena Arens, "Flowerbeds and Hothouses: Botany, Gardens, and the Circulation of Knowledge in Things." *Historical Social Research/Historische Sozialforschung* 40, no.1 (151) (2015): pp.265–83.

Pisa

Gianni Bedini, *The Botanical Garden of Pisa, Plants, history, people, roles*, Pisa University Press, 2019.

Gianni Bedini and Simone Farina, "A Giraffe in the Botanic Garden of Pisa (Tuscany, Northern Italy)." *Der Zoologische Garten* vol.3 no.2 (special issue, April 2022): pp.170–176.

Raffaella Fabiani Gianetto, *Medici Gardens from Making to Design*, Penn, 2008.

https://www.ortomuseobot.sma.unipi.it/

Padua

Margaret Muther D'Evelyn, *Venice and Vitruvius*, Yale, 2012.

The Botanical Garden of Padua, Treasures of Italy and UNESCO, Sagep, 2018.

https://www.ortobotanicopd.it/

https://whc.unesco.org/en/list/824/

Lucia Tongiorgi Tomasi, "Projects for botanical and other gardens: A 16th-century manual," *The Journal of Garden History*, vol.3 no.1: pp.1-35.

Leiden

Gerda van Uffelen, "Hortus Botanicus Leiden." In *SiteLINES: A Journal of Place* 2, no.1 (2006): pp.6-7.

John Dixon Hunt, *Gardens and Grove, the Italian Renaissance Garden in the English Imagination: 1600–1750*, Dent, 1986: note 9 p.238.

https://hortusleiden.nl/en/the-hortus/

Juliet Hodgkiss, "Historic Gardens of the Netherlands," *The Horticulturist* 12, no.2 (2003): pp.9–12.

Montpellier

Charles Frédéric Martins, *Le Jardin des Plantes de Montpellier. Essai historique et descriptif, etc.*, Montpellier, Paris, Strasbourg, 1854.

Charles Frédéric Martins, Opening Speech of the Course of Medical Botany, delivered 17 April 1852, *A Look about History, Botanists and Plant Gardens from Montpellier*, Louis Planghon, printing Ricard Frères, Encivade.

Kenneth Woodbridge, *Princely Gardens, the origins and development of the French formal style*, Rizzoli, 1986.

Albert Fabre, *Histoire de Montpellier: depuis son origine jusqu'a la fin de la révolution avec plusieurs plans de Montpellier (première partie)*, 1897: p.146.

François Michaud, "The revival of the Jardin des Plantes in Montpellier." *Past, present and future of the oldest botanical garden in France* Duo_Jardin_des_plantes_2019_01 (2).pdf

Le Jardin des plantes de Montpellier, patrimoine protegé, monuments historiques et objets d'art d'Occitanie Ministère... see https://www.umontpellier.fr/en/universite/patrimoine/jardin-des-plantes

Copenhagen

Jes Fabricius Møller, "The Parks of Copenhagen 1850–1900." *Garden History* 38, no.1 (2010): pp.112–23.

Oxford

Charles Daubeny, *Oxford Botanic Garden; or a popular guide to the Botanic Garden of Oxford*, printed by I. Shrimpton, 1850.

Stephen A. Harris, *Oxford Botanic Garden & Arboretum, A Brief History*, Bodleian Library, 2017.

Chris Thorogood and Simon Hiscock, *Oxford Botanic Garden, A Guide*, Bodleian Library, 2019.

Anna Svensson, "'And Eden from the Chaos rose': Utopian order and rebellion in the Oxford Physick Garden." *Annals of Science*, vol.76: number 2 (2019, 3 April): pp.157–183.

Karin Seeber, "Jacob Bobart (1596–1680): First Keeper of the Oxford Physic Garden." *Garden History* 41, no.2 (2013): pp.278–284.

Jardin des Plantes Paris

Xavier Riffet, *The Guide Jardin des Plantes*, Muséum National d'Histoire Naturelle, Paris, 2021.

https://www.jardindesplantesdeparis.fr

Amsterdam

D. Onno Wijnands, "The Hortus Medicus Amstelodamensis – Its Role in Shaping Taxonomy and Horticulture." *The Kew Magazine* 4, no. 2 (1987): pp.78–91.

Juliet Hodgkiss, "Historic Gardens of the Netherlands," *The Horticulturist* 12, no. 2 (2003): pp.9–12.

Edinburgh

Harold R Fletcher, William H Brown, *The Royal Botanic Garden Edinburgh 1670–1970*, Her Majesty's Stationery Office, 1970.

Simon Milne, *Director's choice, Royal Botanic Garden Edinburgh*, Scala 2018.

A.G. Morton, *John Hope 1725–1786, Scottish Botanist*, Edinburgh Botanic Garden Trust, 1986.

David Knott, "Horticulture at the Royal Botanic Garden Edinburgh." *The Horticulturist* 21, no.2 (2012): pp.2–5.

4 Gardens in One: Royal Botanic Garden Edinburgh, Edinburgh Stationery Office Books, 1992.

Chelsea Physic Garden

Ruth Stungo, "The Royal Society Specimens from the Chelsea Physic Garden 1722-1799." *Notes and Records of the Royal Society of London*, vol.47, no.2 (1993): pp.213–24.

Sue Minter, *The Apothecaries' Garden: A History of the Chelsea Physic Garden*, The History Press Ltd, 1980.

https://www.chelseaphysicgarden.co.uk/

Madrid

Ricardo R. Austrich, "El Real Jardín Botánico de Madrid and the Glorious History of Botany in Spain." *Arnoldia* 47, no.3 (1987): pp.2–24.

"The Madrid Botanical Garden Today: A Brief Photographic Portfolio." *Arnoldia* 47, no.3 (1987): pp.25–29.

Palmengarten Frankfurt

https://www.palmengarten.de

Sir Seewoosagur Ramgoolam Botanic Garden Mauritius

"The Botanic Garden of Pamplemousses." *Bulletin of Miscellaneous Information* (Royal Botanic Gardens, Kew), vol.1919, no.6/7 (1919): pp.279–86.

Gillian Jones, *Pierre Poivre and the Networking Naturalists, Pioneering Environmentalists of the Eighteenth Century*, Austin Macauley Press, 2022.

Acharya Jagadish Chandra Bose Indian Botanic Garden Calcutta

Adrian P. Thomas, "The Establishment of Calcutta Botanic Garden: Plant Transfer, Science and the East India Company, 1786–1806." *Journal of the Royal Asiatic Society*, vol.16, no.2 (2006): pp.165–177. JSTOR, http://www.jstor.org/stable/25188624

https://bsi.gov.in/garden-page/en?rcu=140,39

Acharya Jagadish Chandra Bose, *Botanical Survey of India*, Indian Botanic Garden, Howrah.

The International Union for Conservation of Nature – for more information on critically endangered species see: https://www.iucn.org/ and https://www.iucnredlist.org/

Arnold, David, "Plant Capitalism and Company Science: The Indian Career of Nathaniel Wallich." *Modern Asian Studies* 42, no.5 (2008): pp.899–928.

Parque de Monserrate Sintra

José da Silva, Gerald Luckhurst, *Sintra, A Landscape with Villas*, Ediçôes Inapa, 1989.

https://www.parquesdesintra.pt/en/parks-monuments/park-and-palace-of-monserrate/

"If mortals were allowed to make their own heaven in their own way, Monserrate would be mine..." *The Gardeners Chronical*, September 1885.

Helder Carita and Homem Cardoso, *Portuguese Gardens*, Antique Collectors Club, 1990.

Jardim Botânico do Rio de Janeiro

Taylor, Norman, "The Botanical Garden at Rio de Janeiro." *Torreya* vol.29, no.2 (1929): pp.25–31.

Royal Botanic Garden Sydney

Fanny Karouta-Manasse, *Discovering Australian Flora*, Stylus Publishing, 2017.

Colleen Morris, *The Florilegium: The Royal Botanic Gardens Sydney – Celebrating 200 Years*, Kew Publishing, 2017.

Charles Moore, *Catalogue of plants in the Government Botanic Gardens, Sydney, New South Wales, 1895*, Leopold Classic Library, 2015.

https://www.rbgsyd.nsw.gov.au/

Tresco

Mike Nelhams, *Tresco Abbey Garden: A Personal and Pictorial History*, Truran, 2000.

Paula Deitz, "Tresco Abbey Garden." *SiteLINES: A Journal of Place*, vol.8, no.1 (Fall 2012): pp.4–7.

"A visit to S.W. Cornwall and the Scillies: where cacti and succulents grow in the open air." *The Cactus and Succulent Journal of Great Britain*, vol.8, no.4 (October 1946): pp.88–89.

R. Senior, "Succulent plants outdoors in Britain No. 2, Tresco." *The National Cactus and Succulent Journal* 23, no.4 (1968): pp.87–89.

Kew Gardens

Kate Teltscher, *Palace of Palms, Tropical Dreams and the Making of Kew*, Picador, 2020.

Patricia Fara, *Sex, Botany & Empire, The Story of Carl Linnaeus and Joseph Banks*, Icon, 2003.

Lucile H Brockway, "Science and Colonial Expansion: The Role of the British Royal Botanic Gardens." *American Ethnologist* 6, no. 3 (1979): pp.449–65.

Mark Nesbitt and Caroline Cornish, "Seeds of Industry and Empire: Economic Botany Collections between Nature and Culture." *Journal of Museum Ethnography*, no. 29 (2016): pp.53–70.

Royal Botanic Gardens Melbourne

R T M Pescott, *The Royal Botanic Gardens, Melbourne: A History from 1845 to 1970*, Oxford University Press, 1982.

Richard Aitken and William R. Guilfoyle, "William Guilfoyle's First Decade at the Melbourne Botanic Gardens." *Australian Garden History* 7, no. 5 (1996): pp.6–18.

Joan Law-Smith, "Rejuvenating Melbourne's Royal Botanic Gardens." *Australian Garden History* 5, no. 3 (1993): pp.3–5.

Georgina Whitehead, "The Influence of Environmental Thought in Melbourne's Nineteenth-Century Public Gardens." *Australian Garden History* 20, no.1 (2008): pp.10–17.

Missouri Botanical Garden

"The Missouri Botanical Garden." *Science* 11, no.266 (1900): pp.196–97.

https://www.missouribotanicalgarden.org/

The Botanic Gardens of Singapore

John Bastin, "The Letters of Sir Stamford Raffles to Nathaniel Wallich 1819–1824." *Journal of the Malaysian Branch of the Royal Asiatic Society* 54, no.2 (240) (1981): pp.1–73.

Pehr Olsson-Seffer, "Visits to some Botanic Gardens abroad (continuation)." *The Plant World* 10, no.6 (1907): pp.130–37.

Tan Wee Kiat, "Keeping Botanical Gardens Relevant – The Singapore Botanic Gardens Experience." *Botanic Gardens Conservation News* 3, no.3 (1999): pp.45–48.

Bagh-e-Jinnah Lahore, Pakistan

Abdul Rehman, "Changing concepts of garden design in Lahore from Mughal to contemporary times." *Garden History*, Winter 2009, vol.37, no.2 (Winter 2009): pp.205–217.

Jardín Botánico Historico La Concepción, Malaga

Blanca Lasso de la Vega, Amelia Denis, Belén Verdú and Carlos Espejo, *Jardín Botánico Histórico La Concepción Malaga Official Guidebook*, Malaga City Council, 2018.

Hanbury Garden La Mortola

Alasdair Moore, *La Mortola in the footsteps of Thomas Hanbury*, Cadogan, 2004.

Hanbury Botanic Garden Guide by the Università degli Studi di Genova.

Penang

https://botanicalgardens.penang.gov.my

R E Holttum, *The Waterfall Garden, Penang, Illustrated Guide* (1934), see: https://botanicalgardens.penang.gov.my/images/The%20Waterfall%20Garden%20Penang.pdf

New York Botanic Garden

Victoria Johnson, *American Eden, David Hosack, Botany, and Medicine in the Garden of the Early Republic*, Liveright, 2018.

Gerry Moore, Angela Steward, Steven Clemants, Steven Glenn, and Jinshuang Ma, *An Overview of the New York Metropolitan Flora Project* https://www.urbanhabitats.org/v01n01/nymf_full.html

Gregory Long, Anne Skillion (ed.), *The New York Botanical Garden*, Abrahams, 2006.

https://www.nybg.org/

Dublin

E. Charles and Eileen M McCracken, *The Brightest Jewel, A History of the National Botanic Gardens Glasnevin*, Dublin, Boethius Press, 1987.

https://www.botanicgardens.ie/

Berlin

E. B. Babcock, "The Berlin Botanic Garden." *Science New Series*, vol.107, no.2789 (June 1948), p.622. Published by the American Association for the Advancement of Science.

Katja Kaiser, "Exploration and exploitation: German colonial botany at the Botanic Garden and Botanical Museum Berlin." In *Sites of Imperial Memory: Commemorating Colonial Rule in the Nineteenth and Twentieth Centuries*, ed. by Dominik Geppert and Frank Lorenz Müller, Manchester University Press (2015): pp.225–42.

Katja Kaiser, "Duplicate networks: the Berlin botanical institutions as a 'clearing house' for colonial plant material, 1891–1920." *The British Journal for the History of Science*, 55 (3): pp.279–296.

https://www.bgbm.org/en/historical-background

Brooklyn

Gerry Moore, Angela Steward, Steven Clemants, Steven Glenn, and Jinshuang Ma, *An Overview of the New York Metropolitan Flora Project* https://www.urbanhabitats.org/v01n01/nymf_full.html

Sally Gregory Kohlstedt, "Mobile Botany: Education, Horticulture and Commerce in New York Botanical Garden, 1890s–1930s." In *Mobile Museums: Collections in Circulation*, ed. Felix Driver, Mark Nesbitt, and Caroline Cornish, UCL Press (2021): pp.178–205.

Official Website of the New York City Department of Parks & Recreation: https://www.nycgovparks.org/parks/brooklyn-botanic-garden/history

Kirstenbosch National Botanical Garden

Elizabeth Green Musselman, "Plant Knowledge at the Cape: A Study in African and European Collaboration." *International Journal of African Historical Studies* vol.36, no.2 (2003): pp.367–92. https://doi.org/10.2307/3559388

Robert Harold Compton, *Kirstenbosch, Garden for a Nation*, Tafelberg-Uitgewers, 1965.

Brian J Huntley, *Kirstenbosch: The Most Beautiful Garden in Africa*, Struik Nature, 2012.

"Kirstenbosch." *The Cactus Journal* 4, no.1 (1935): p.18.

Huntington

Walter Houk, *The Botanical Garden at the Huntington*, Huntington Library Press, 1996.

"The Huntington Botanical Garden." *The Cactus Journal* 4, no. 1 (1935): p.16.

Paula Panich, "California Treasure: Henry Huntington's Jewel-Box Desert." *SiteLINES: A Journal of Place* 10, no.2 (2015): pp.12–14.

T. June Li, "Creating Liu Fang Yuan, the 'Garden of Flowing Fragrance' in California." *Journal of the Royal Asiatic Society Hong Kong Branch* 55 (2015): pp.183–98.

Botaniska Trädgård Gothenburg, Sweden

Magnus Lidén, "The Genus 'Dionysia (Primulaceae)', a Synopsis and Five New Species." *Willdenowia*, vol.37, no.1 (2007): pp.37–61.

https://www.botaniska.se/en/garden--greenhouses/greenhouses/easter-island-tree/

Majorelle

Pierre Bergé and Madison Cox, *Majorelle: A Moroccan Oasis*, Thames and Hudson, 1999.

Manon Garrigues (translated by Hannah Larvin), "Take a virtual tour of the Majorelle Garden, a centre of peace and serenity in Marrakesh." *French Vogue*, May 2020.

https://majorelle-gardens.com/

https://www.museeyslmarrakech.com/en/fondation-jardin-majorelle/le-jardin-majorelle/

Kyoto Shokubutsuen

Jitin Rahul et al., "Adansonia digitata L. (baobab): a review of traditional information and taxonomic description." *Asian Pacific Journal of Tropical Biomedicine*, vol.5, issue 1 (January 2015): pp.79-84.

Fairchild

Mike Maunder, "Fairchild Tropical Botanic Garden." *SiteLINES: A Journal of Place*, vol.2, no.1, 2006.

https://fairchildgarden.org/mission-history/

https://news.artnet.com/art-world/chihuly-returns-to-tropical-botanical-garden-201261

Australia Mount Annan

Research project *ARC Linkage (2019–22)*. The Australian National University, Research School of Biology, ANU College of Science.

https://www.dezeen.com/2014/06/12/australian-plant-bank-bvn-donovan-hill-seed-preservation/

https://www.australianbotanicgarden.com.au/

Tromsø Arctic-Alpine Botanical Garden

David W H Rankin, "726. PRIMULA MELANANTHA: Primulaceae." *Curtis's Botanical Magazine* 29, no.1 (April 2012): pp.18–33.

https://en.uit.no/tmu/botanisk

First published in Great Britain in 2023 by Greenfinch
An imprint of Quercus Editions Ltd
Carmelite House
50 Victoria Embankment
London
EC4Y 0DZ

An Hachette UK company

A CIP catalogue record for this book is available from
the British Library.

HB ISBN 978-1-52942-809-4
eBook ISBN 978-1-52942-810-0

10 9 8 7 6 5 4 3 2 1

Cover design by Sarah Pyke
Internal design by Paul Turner and Sue Pressley,
Stonecastle Graphics
Edited by Nick Freeth

Printed and bound in Italy

Papers used by Greenfinch are from well-managed
forests and other responsible sources.